Introduction to

MACROMOLECULAR BINDING EQUILIBRIA

Introduction to
MACROMOLECULAR
BINDING EQUILIBRIA

Charles P. Woodbury

CRC Press
Taylor & Francis Group
Boca Raton London New York

CRC Press is an imprint of the
Taylor & Francis Group, an **informa** business

CRC Press
Taylor & Francis Group
6000 Broken Sound Parkway NW, Suite 300
Boca Raton, FL 33487-2742

First issued in paperback 2019

ISBN-13: 978-1-4200-5298-5 (hbk)
ISBN-13: 978-0-367-38832-4 (pbk)

Library of Congress Cataloging-in-Publication Data

Woodbury, Charles P.
    Introduction to macromolecular binding equilibria / author, Charles P. Woodbury Jr.
    p. ; cm.
    "A CRC title."
    Includes bibliographical references and index.
    ISBN-13: 978-1-4200-5298-5 (hardcover : alk. paper)
    ISBN-10: 1-4200-5298-5 (hardcover : alk. paper)
    1. Binding sites (Biochemistry) 2. Ligand binding (Biochemistry) 3.
Macromolecules. I. Title.
    [DNLM: 1. Binding Sites. 2. Ligands. 3. Macromolecular Substances. 4.
Models, Molecular. QU 34 W884i 2008]

QP517.B42W66 2008
572'.33--dc22
                                        2007020733

Visit the Taylor & Francis Web site at
http://www.taylorandfrancis.com

and the CRC Press Web site at
http://www.crcpress.com

Dedication

To my beloved aunt, Martha Ann Woodbury.
Thank you for sharing your love of books.

Contents

Preface

This book is based on my experience in teaching biophysical chemistry to graduate students in the medicinal chemistry, pharmaceutics, and bioengineering programs at the University of Illinois at Chicago. The course was designed to give these students an adequate background in biophysical chemistry for research in drug discovery and development, and the course has had a strong emphasis on macromolecular solvation and ligand binding. The role of hydration in binding phenomena, along with the effects of salt and pH, is increasingly emphasized in the literature, so these students need a somewhat deeper background in preferential interaction concepts and in linkage thermodynamics than might students in biochemistry, pharmacology, or physiology programs. Of course, the students also need to learn about new techniques for characterizing macromolecular complexes with small molecules, such as surface plasmon resonance and fluorescence polarization, but there are certain underlying concepts in binding theory they need to grasp firmly so that they can interpret the results of assays that use these new methods.

This book tries to give that necessary theoretical background. It is certainly not a complete exposition of macromolecular binding, only an introduction that emphasizes some selected fundamental topics. The analytical methods of assaying macromolecular binding I have mainly left aside, though I do present what I hope will be some useful general advice on designing binding assays and how to interpret them. I also have summarized qualitative features of binding sites on proteins and nucleic acids, an area that is now referred to as "molecular recognition". But my main purposes have been to show how to use the binding polynomial approach in model building and interpretation, and to show how linkage thermodynamics can tie together disparate binding observations. Whenever possible, I have tried to tie the theory to concrete examples drawn from the research literature. I have also included a substantial number of references to the original literature, for those who might wish to pursue further any of the topics presented.

I would like to acknowledge the support given me by the Department of Medicinal Chemistry and Pharmacognosy at the University of Illinois at Chicago, and by my colleagues in that department. And of course, I welcome comments and suggestions from readers for improvement of this book.

Charles Woodbury
University of Illinois at Chicago

Acknowledgments

There have been many sources of inspiration for this book. First, my students, who endured a biophysical chemistry course full of messy handouts and photocopied journal articles. This book represents what I wanted to say to you with those handouts and articles. Sorry it took a little longer than expected.

Second, my mentors: Tom Record and Pete von Hippel. You put me on track, and I thank you for your patience, support, and encouragement, when it really counted.

And lastly, speaking of patience, support, and encouragement, I must thank my wife Marty, for putting up with a grouch who spent entirely too much time in front of a keyboard, muttering to himself "isotherm," "linkage," "cooperativity," and similar gibberish. Let's take a vacation!

The Author

Charles Woodbury was born in 1949 in El Paso, Texas. Raised in Virginia and the state of Washington, he attended the University of Washington in Seattle for an undergraduate degree (B.S.) in chemistry in 1971, then went on to finish a Ph.D. in physical chemistry from the University of Wisconsin (Madison) in the fall of 1975, under Professor Tom Record. After a very enjoyable postdoctoral stint at the University of Oregon in the Institute for Molecular Biology, in Professor Peter von Hippel's research group, he joined the Department of Medicinal Chemistry (later renamed the Department of Medicinal Chemistry and Pharmacognosy) at the University of Illinois Medical Center in Chicago as an assistant professor. He is currently an associate professor in that department. His research interests lie in the area of biophysical chemistry, including polyelectrolytes, macromolecular binding and recognition, and theory and analytical applications of chromatography.

1 Binding Sites

> Except when radiation participates, all biological activities involve contact interactions between constituent reactants.
>
> **Irving M. Klotz (1985)** *Q. Rev. Biophys.* **18, 227–259**

1.1 THE IMPORTANCE AND COMPLEXITY OF MACROMOLECULAR BINDING

Reversible, noncovalent associations are fundamental to biochemistry. Noncovalent associations control gene expression, regulate metabolism, pass signals across membranes and between cells, and enable the body to recognize and reject invaders like viruses and bacteria. Often, the association involves a small molecule or ion, and a much larger partner, such as a protein or nucleic acid, but contact associations between two or more macromolecules (e.g., DNA-protein or protein-protein complexes) are frequently just as important. Changes in the concentration of one of the partners may lead to only modest alterations in cellular functions and responses—for example, a slight increase in cellular respiration as ATP is consumed in mild exercise—or the change may provoke quite drastic responses, such as the anaphylactic shock reaction to a bee sting, for someone with hypersensitivity to bee venom. To understand the origins, magnitudes, and range of such responses, it is necessary to have a firm grasp of the essentials of the underlying macromolecular binding equilibria. To provide those essentials is the aim of this book.

1.1.1 DIFFERENT TYPES OF MULTIPLE EQUILIBRIA IN MACROMOLECULAR BINDING

Macromolecular binding typically involves *multiple equilibria*, and these equilibria can be of several kinds. A short list of these equilibria could start with the aggregation or dissociation of subunits of macromolecules; the list might then include changes in the conformation of the macromolecule that affect its ability to bind ligands; and it should certainly note the possibility for a single macromolecule to have multiple sites for binding a given species of ligand. As a result, there may be several different subspecies of ligand-macromolecule complexes in equilibrium with each other. Figure 1.1 illustrates some of these complexities for a familiar system, the oxygenation of hemoglobin. The hemoglobin tetramer is in equilibrium with heterodimers; the individual protein subunits can change conformation, an important factor in their affinity for oxygen; the uptake of one ligand species, O_2, is influenced by (linked to) the presence of a second small molecule species, H^+, in the well-known Bohr effect; and the oxygenation process involves cooperative interactions among the protein subunits, tied to the conformational changes in those subunits.

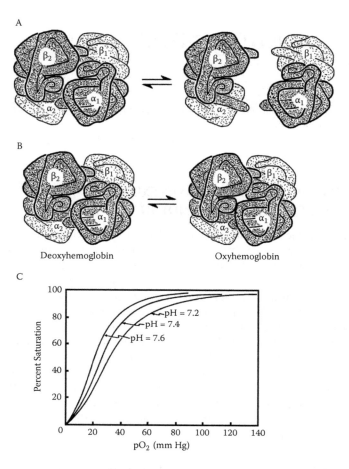

FIGURE 1.1 The complexities of macromolecular binding as exemplified by the oxygenation of hemoglobin. A. A highly schematized representation of the hemoglobin molecule as a tetramer with two kinds of subunit, α and β, with the stoichiometry $\alpha_2\beta_2$. The tetramer is in equilibrium with $\alpha\beta$ dimers, with the tetramer favored under physiological conditions. The tetramer has four heme groups with which to bind oxygen and there are multiple possible states of oxygenation of the hemoglobin molecule. In the figure, numerical subscripts on α and β indicate subunit origin with respect to the dimer species. B. Oxygenation of the tetramer involves conformational changes in the subunits, with these changes propagated across the interfaces between subunits. Compared to the deoxygenated form, oxygenated hemoglobin has a smaller central cavity and different contacts across the α_2 and β_1 chains, and across the α_1 and β_2 chains. C. The degree of oxygen saturation of the hemoglobin depends on the solution pH; protonation equilibria change the affinity of hemoglobin for O_2, and link hemoglobin's binding behavior to local physiological conditions. The oxygenation curve for hemoglobin is not a simple hyperbolic curve but is instead sigmoid in shape; this is characteristic of cooperative interactions among the protein's subunits that affect their affinity for O_2. Adapted from *Advances in Protein Chemistry* Vol. 28, R.E. Benesch and R. Benesch, "The mechanism of interaction of red cell organic phosphates with hemoglobin," pages 211–237. Copyright 1974, with permission from Elsevier.

The phenomenon of *binding linkage*, in which the uptake of one kind of ligand increases or decreases the binding of a different ligand species, is widespread throughout biochemistry [1,2]. Familiar examples here include pH effects on enzyme kinetics, and the Bohr effect in the pH dependence of the oxygenation of hemoglobin; the oxygenation of hemoglobin is also sensitive to the presence of the small organic molecule 2,3-bisphosphoglycerate, which can substantially lower hemoglobin's affinity for oxygen. Small inorganic ions, such as chloride or sodium ions, are frequently linked to macromolecular equilibria, e.g., protein or nucleic acid conformational changes, protein-protein aggregation, inhibitor binding by enzymes, etc.; also, protein-nucleic acid interactions are well known for their sensitivity to salt concentration [3]. The solvent water can also participate in binding equilibria. There can be large changes in hydration of a macromolecule across a binding equilibrium, and the degree of hydration may play a significant role in modulating the binding equilibrium in vivo [4,5].

Cooperativity is often a feature in macromolecular binding systems [2]. A classic example is the oxygenation of hemoglobin, in which there is minimal uptake of oxygen until a certain threshold partial pressure of O_2 is reached, and then nearly full oxygenation is reached with only a small further increase in O_2 partial pressure. This behavior is described as *positive cooperativity*. Positive cooperativity in the oxygenation of hemoglobin provides for loading and unloading of oxygen over a small range of oxygen partial pressures, e.g., the conditions obtained in the lung versus those in working muscle tissue. Positive cooperativity is also found with numerous enzymes, where small changes in the concentration of a small molecule effector can drastically enhance the enzyme's catalytic activity. Generally speaking, positive cooperativity increases the sensitivity of a biochemical system to local conditions, to produce large changes in activity or function in response to relatively small changes in the concentration of the ligand. In contrast, *negative cooperativity* describes a situation in which the initial event (binding of a ligand species, binding of substrate, etc.) dampens or reduces a further event of the same type (uptake of yet more ligand or substrate, etc.). This kind of cooperativity is often harder to discern than is positive cooperativity. Instances can be found, however, in the binding of some drugs and dye molecules to DNA, and in the binding of cofactors by certain enzymes.

These processes all take place in an aqueous environment. Water, of course, participates in numerous biochemical reactions as either a reactant or a product, where covalent interactions are broken and formed. But water's noncovalent interactions with solutes are just as important. Water is well known for its ability to dissolve a wide range of substances; the aqueous solvation of solutes both large and small is done through weak, noncovalent interactions. Water is, however, a very self-cohesive liquid, which leads to the notable lack of solubility of nonpolar substances in water. This hydrophobic effect plays a major role in the stabilization of macromolecular assemblies as well as in the binding of small molecules by biopolymers [6]. Beyond this, water often participates directly in the fitting together and stabilization of macromolecular surfaces, by filling cavities and by linking molecular surfaces through hydrogen bonding. Finally, there is competition for access to the surfaces of biopolymers between water and small solutes or ions, and this is an important means of thermodynamically linking biochemical processes, for efficient communication and regulation.

In this book there will be practically no details of experimental methods, on how to measure the fluorescence change of a protein as it binds a ligand, or how the electrophoretic mobility of a nucleic acid may change as it takes up an intercalating dye. Likewise, there will be little here on the kinetics of ligand binding. Instead, after a survey of the characteristics of binding sites, this book will focus primarily on commonly used equilibrium models for macromolecular binding, explaining details of the models and showing how they have been applied to real systems. It will first treat simple binding equilibria, in which there is one species of macromolecule and one species of ligand. In subsequent chapters it will explore the effects of binding linkage and of binding cooperativity in more complicated binding systems. It will also consider the effects of forming binding sites into a linear array, as with DNA and other linear biopolymers, where ligands make contact with more than one site at a time. Finally, it will offer some suggestions on designing binding experiments and interpreting the data, general approaches that can be used regardless of the specific experimental technique used to generate the data.

1.2 GENERATING AFFINITY AND SPECIFICITY WITH WEAK INTERACTIONS

1.2.1 WEAK INTERACTIONS AND REVERSIBLE BINDING

The types of interactions that are responsible for holding a ligand in complex with a macromolecule are the same as those for holding the macromolecule in a particular conformation. Included here are hydrogen bonds, dispersion and exchange interactions (which together with polarization interactions are commonly collected under the name of van der Waals interactions), and ionic and other electrostatic interactions.

These interactions are energetically weak, on par with the kinetic energy of thermal agitation, which is about 2.5 kJ/mol at room temperature. In the vapor phase some of these interactions may be considerably stronger, especially hydrogen bonds, because there is no longer any direct competition with surrounding water molecules that could serve as alternative bonding partners; also, the high dielectric constant of an aqueous solution considerably moderates electrostatic interactions, compared to the vapor phase [7,8]. Simple thermal agitation can readily disrupt individual weak interactions, and the simultaneous breaking of two or more interactions is less likely but still possible. While the typical ligand-macromolecule complex will have several of these weak interactions operating simultaneously, it is nevertheless possible to break up the complex by simple thermal agitation; that is, the binding/recognition process is usually reversible. If covalent bonds were used instead, the binding would be essentially irreversible. The reversibility of these interactions is important for biological functions. If binding were always irreversible, an enzyme would never release its substrate, and a receptor would stay occupied by agonist (or antagonist) forever.

While breakage of a single weak interaction can occur very quickly, the simultaneous breakage or disruption of multiple weak interactions, as needed to release a ligand, will likely not occur so quickly. The lifetime of a ligand-macromolecule complex depends on the multiplicity as well as the strength of these interactions. The duration of such complexes may be important in biological regulation, for example in signal transduction and gene regulation, in which responses might need to be delayed in order to coordinate different systems, or to avoid responding too dramatically to sharp but transient changes in the environment.

1.2.2 BINDING SPECIFICITY AND MULTIPLE SIMULTANEOUS WEAK INTERACTIONS

Macromolecular systems often exhibit a high degree of *specificity* in binding ligands. The term *specific* as applied to binding implies the rejection of incorrect binding partners in favor of the correct ones, as measured in terms of a ratio of equilibrium binding constants, or the binding constant of the "correct" partner divided by that of the "incorrect" partner [9]. High specificity sharpens biological responses, as in the functioning of the immune system: antibodies recognize antigens with exquisite specificity, and discriminate among possible antigens that differ in only the most minor of chemical details. Specificity also promotes fidelity, the faithful performance of the same (biological) action time after time, as in the faithful replication of DNA by DNA polymerase.

The strength and directionality of the weak interactions involved in ligand-macromolecule complexes are critically important for binding specificity. At its base, binding specificity derives from the highly organized structure of the functional groups around the binding site in a macromolecule, and the three-dimensional alignment of these groups with complementary features on a prospective ligand. Binding involves the coming together of two molecular surfaces, that of the ligand and that of the macromolecule. For there to be specificity in the binding, the two molecular surfaces should be complementary to each other (Figure 1.2); that is, they should

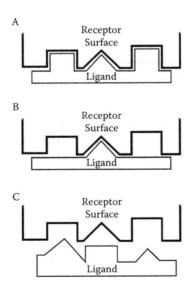

FIGURE 1.2 Close steric fit is important for high affinity and specificity in macromolecular binding. Gaps or voids in the receptor-ligand interface will reduce affinity and affect specificity. A. Close match of complementary molecular surfaces results in high affinity. B. Partial match of molecular surfaces results in lower affinity. C. Lack of matching molecular surfaces results in little to no binding affinity.

fit together sterically, without substantial voids or gaps in the interface. Furthermore, there should be alignment of functional groups across the interface that can establish attractive noncovalent interactions (e.g., hydrogen bonds and electrostatic interactions; see Figure 1.3). Typically, there will be multiple simultaneous noncovalent interactions between ligand and macromolecule, creating a three-dimensional network of contacts between macromolecule and ligand. Voids at the interface will reduce the overall van der Waals attraction between the two surfaces and so are disfavored, and misaligned steric features, hydrogen bonds, ion pairs, etc., can seriously destabilize the complex.

Because weak forces are involved, macromolecular binding events typically have much smaller enthalpy and free energy changes than those found with covalent bond formation. As a result, a macromolecular binding reaction often does not go to completion. Instead, the situation resembles the titration of a weak acid or base, with the gradual equilibrium formation of products (ligand-macromolecule complexes) upon increasing the amount of one component or the other. The incompleteness of the binding reaction and the typically rapid exchange of material across the equilibrium tend to make it difficult to define experimentally the stoichiometry and detailed structure of such weak complexes. At the same time, however, the characteristic quality of the partial or less-than-stoichiometric reaction permits shifting the equilibrium in response to small changes in conditions, a sensitivity that is often quite desirable from a biological point of view.

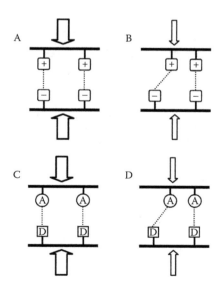

FIGURE 1.3 Proper alignment of points of weak interactions will lead to high binding affinity, while misalignment of contacts will lower the binding affinity. A. Strong, aligned ionic attractions. B. Perturbed, weakened ionic interactions. C. Strong, aligned hydrogen bond donors and acceptors (A = acceptor, D = donor). D. Distorted, weakened hydrogen bonds.

1.2.3 THE STRENGTH OF BINDING

The discussion must start with the fundamental connection of the equilibrium constant to the free energy change: $\Delta G° = -RT \ln K$. Typical binding (association) constants for specific drug-receptor and enzyme-inhibitor interactions are in the range $K = 10^4$ M^{-1} to $K = 10^{11}$ M^{-1}. The strongest known noncovalent binding affinity for a biochemical system is that of streptavidin for biotin, with a binding constant on the order of 10^{15} M^{-1} [10]. This can be compared to the binding between proteins and urea (a common denaturant), which is very weak with K of the order of 1 M^{-1} [11].

To put this into perspective, to achieve a binding constant of 10^6 M^{-1} (which is moderately strong binding, typical of an initial "hit" in a drug discovery program, for example), the free energy change ΔG at room temperature must be about 34 kJ/mol. For a binding constant of 10^9 M^{-1} (very strong binding), ΔG must be about 50 kJ/mol. Assuming an average value of 4 to 8 kJ/mol for a single noncovalent contact, this implies that four or more contacts are needed to stabilize these complexes, even those of modest stability.

Knowledge of $\Delta G°$ alone is not sufficient to characterize the binding mechanism. $\Delta G°$ has both enthalpic and entropic contributions: $\Delta G° = \Delta H° - T \Delta S°$. For more insight into the origin of binding specificity and affinity in a system, it is also necessary to determine the values of $\Delta H°$ and $\Delta S°$. The magnitude of $\Delta H°$ can help indicate the relative contributions of hydrogen bonding, multipolar force, dispersive interactions, etc., while the magnitude of $\Delta S°$ can help indicate the role of solvent reorganization, and possibly internal rigidification or flexibility.

1.2.4 ENTHALPY-ENTROPY COMPENSATION

A number of biologically important processes, including protein folding, solvation of nonpolar compounds, and the binding of ligands to macromolecular receptors, show a striking thermodynamic phenomenon, that of *enthalpy-entropy compensation* [12,13]. When pairs of values of $\Delta H°$ and $\Delta S°$ for these processes are plotted, it appears that the enthalpy change is a linearly increasing function of the entropy change; that is, $\Delta H° = \alpha + T_c \, \Delta S°$, with α a constant for a given set of solvent conditions, and T_c a constant that is positive. The phenomenon is not limited to equilibrium processes; it also appears in certain chemical kinetic processes, in which the linear relation now is between the activation enthalpy change and the activation entropy change. In either case, the slope T_c has the units of temperature, and T_c is called the compensation temperature.

In some cases, the compensation in nearly complete; that is, the factor α is small, negligible by comparison to either $\Delta H°$ or $T_c \, \Delta S°$. Experimentally, enthalpy-entropy compensation is most obvious when it is complete, so that $\Delta G°$ does not change despite significant correlated changes in $\Delta H°$ and $T \, \Delta S°$. But quite commonly the compensation is only partial, especially with complicated systems like biopolymeric drug receptors. While there may still be a linear correlation of $\Delta H°$ with $\Delta S°$, with partial compensation the free energy change $\Delta G°$ will vary (perhaps erratically or perhaps systematically) across the series of compounds.

Part of the difficulty in doing accurate thermodynamic work on drug-receptor binding is obtaining sufficient quantities of purified receptor in soluble form, especially for calorimetry. This is particularly important for membrane-bound receptors, since these are usually present in biological tissues at very low concentrations (estimated, for example, at 1 to 100 fmol/mg tissue for neurotransmitter receptors). The low concentrations do not permit calorimetric studies of binding, so instead the temperature dependence of the binding constant is analyzed through van't Hoff plots of the logarithm of the equilibrium binding constant as a function of reciprocal temperature. Such plots are often linear when the ligand is a small molecule, at least over the limited temperature ranges typically studied. It then appears that the change in heat capacity ΔC_P° is small. Under these conditions the van't Hoff plots can be analyzed to yield values for ΔH° and ΔS°.

Gilli et al. [14] surveyed a variety of drug-receptor systems and found there was a strong linear correlation of ΔH° with ΔS°:

$$\Delta H^\circ \text{ (kJ/mol)} = -39.8\,(\pm0.8) + 278\,(\pm4)\Delta S^\circ \text{ (kJ/mol)} \qquad (1.1)$$

A total of 186 pairs of values, from 13 different receptor systems and with 136 different ligands, were correlated, with a remarkably high correlation coefficient r of 0.981. The data set included not only drug binding to membrane-bound receptors but also inhibitor binding to soluble enzymes (renin and dihydrofolate reductase) and anthraquinone binding to DNA. A recent update of the data set to include 436 drug-receptor pairs, for membrane-bound receptors only, gave again a very strong linear correlation of $T\Delta S^\circ$ with ΔH° [15]. See Figure 1.4.

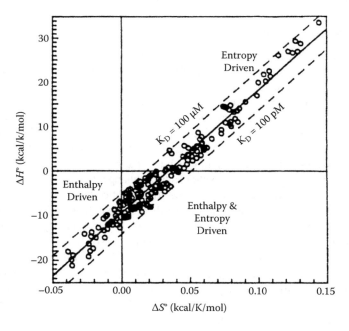

FIGURE 1.4 Correlating ΔH° with ΔS° for a variety of receptor-ligand pairs. Adapted with permission from *Journal of Physical Chemistry* Vol. 98, P. Gilli et al., "Enthalpy-entropy compensation in drug-receptor binding," pages 1515–1518. Copyright 1994 American Chemical Society.

The correlation may also be written as

$$\Delta H^\circ - 278\,\Delta S^\circ = -39.7 \text{ kJ/mol} = \Delta G^\circ \qquad (1.2)$$

This can be interpreted as a sort of average value for the binding free energy change for the data surveyed. The numerical value for ΔG° corresponds to a binding constant of about 1×10^7 M^{-1}; Figure 1.4 shows this as the central heavy continuous line through the data. The figure also contains two dashed lines that form a band within which the $\Delta H^\circ/\Delta S^\circ$ data pairs fall. The lines mark two extremes in affinity, binding constants of 10^{11} M^{-1} and 10^4 M^{-1}, which correspond to binding free energy changes of -63 kJ/mol and -23 kJ/mol, respectively. The upper line marks the boundary for low affinity binding, and the lack of data beyond this boundary is really an artifact of medicinal chemists and pharmacologists having selected only high affinity ligands to study. Drugs with binding constants below 10^4 M^{-1} would probably not be studied in the first place. The lower line at -63 kJ/mol (corresponding to $K = 10^{11}$ M^{-1}) is a different sort of boundary, however, possibly set by the nature of ligand-receptor interactions, or possibly by the lack of sensitive techniques to detect binding in the dilute solutions needed for such high binding constants.

Is the correlation between the changes in enthalpy and entropy real (that is, due to underlying chemical cause), or is it the result of error propagation in evaluating entropy changes from van't Hoff plots? It may well happen that experimental errors in the free energy change (that is, in ln K) are much smaller than errors in the enthalpy change. Then, since the value for ΔS is determined from the relation $\Delta S = (\Delta H - \Delta G)/T$, the error in ΔS will mostly reflect (be correlated with) the error in ΔH, and will lead to a spurious correlation of the (graphically estimated) values of ΔS with ΔH. A plot of ΔH as a function of $T\Delta S$ may also be seriously misleading [16].

Krug and coworkers [17–19] have suggested two criteria that are sufficient, but not necessary, to support a chemical cause for the correlation over an accidental correlation through error propagation in a van't Hoff analysis: first, the compensation temperature T_c must be significantly different from the experimental temperature; and second, values of ΔH° should be linearly correlated with values of ΔG°. Liu and Guo [13] suggest instead an examination of a plot of ΔS versus ΔH that includes error bars in both dimensions; a linear correlation of ΔS with ΔH is highly questionable if there is substantial overlap of errors across the plot. In many cases, it appears that the reported correlations are indeed spurious [20,21].

The results from Gilli et al. do not meet the criteria set forth by Krug et al. However, the same general trend in enthalpy-entropy compensation occurs with several soluble enzyme systems studied by microcalorimetry, e.g., RNase with 3´-CMP and dihydrofolate reductase with various inhibitors; it is also seen with drug-DNA binding equilibria and with carbohydrate-protein binding equilibria. The use of calorimetry to obtain the enthalpy change, instead of a van't Hoff plot, sidesteps the problem of accidental correlation of ΔH° with ΔS°. Further detailed analysis of the data for propagation of error supports the nonartifactual explanation for the correlation found by Gilli et al.

What is the chemical basis for enthalpy-entropy compensation? The simplest explanation is that when ligand and receptor interact strongly, having a large and negative $\Delta H°$, the resulting complex may well be rigid, resulting in a negative $\Delta S°$. Weaker complexes would presumably be less rigid and more flexible, with less of an entropic cost for their formation. Two more sources of compensation should be considered, however: (1) solvent reorganization [13,22]; and (2) isomerization equilibria, with "isomerization" taken in a general sense that includes conformational changes, aggregational equilibria, and so on. These can lead to a change in heat capacity across the equilibrium, $\Delta C_P°$, which engenders compensating changes in $\Delta H°$ with $\Delta S°$ [23]. In fact, basic statistical thermodynamic considerations indicate that, in general, whenever there is an equilibrium dominated by multiple weak intermolecular interactions, there will be an appreciable $\Delta C_P°$ and hence, enthalpy-entropy compensation will appear (at least partially) across that equilibrium [24–27].

1.3 SIZE, SHAPE, AND FUNCTIONAL COMPLEMENTARITY DETERMINE RECOGNITION

1.3.1 EXPOSED SURFACES AND BINDING

Usually, it is not the interior of a macromolecule that is recognized by another molecule. Instead, recognition involves the contact of one exposed surface with another. Aggregation, enzyme action, receptor activation/inhibition, etc., all involve forming contacts between the surfaces of two different molecules. There are exceptions, of course, in which a small molecule may be buried in the interior of a larger molecule, but there is still contact between molecular surfaces.

In the separate (unbound) molecules these surfaces are exposed to solvent, and this solvent often must be removed in order for a complex to form. (The area exposed to solvent by a molecule, of course, plays a significant role in its solubility.) The release of the bound solvent is usually favorable in terms of entropy, and so may help in driving complex formation. Some solvent molecules may remain in the binding site, where they can serve to fill in vacant spaces, help to form hydrogen bond networks, and possibly serve as a molecular-level lubricant for conformational changes.

1.3.1.1 Accessible Surface Area

A key quantity here is the accessible (solvent-exposed) surface area, abbreviated as ASA. The ASA is quantitatively determined by rolling a virtual ball (a probe) over the outside of the macromolecule of interest. The size of the ball is usually chosen to mimic a water molecule, in order to detect and quantify clefts and pockets in the surface where water could be expected to penetrate. Typically, the radius of the probe is set to 1.4 Å, based on the 2.8 Å distance between nearest-neighbor oxygen atoms in ice crystals, a representative diameter for a water molecule. The atoms of the macromolecule are modeled as hard spheres with appropriate radii (see Figure 1.5). Of course, to perform the calculation it is necessary to have the atomic coordinates of the macromolecule's structure. Fortunately, these are often readily available through the Internet from the Protein Data Bank (PDB) and the Nucleic Acid Database (NDB), curated archives of crystal structures for biomacromolecules [28–32].

The ASA naturally depends on the macromolecule's conformation. Native proteins are tightly folded and will present much less surface to solvent than would the

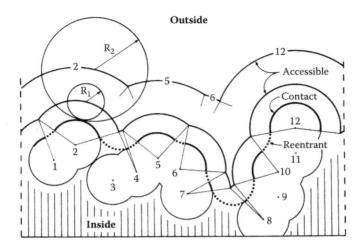

FIGURE 1.5 The accessible surface of a biopolymer, as perceived by two probes of different radii. Note how the larger probe rolls over crevices that the smaller probe will penetrate. Reprinted, with permission, from *Annual Review of Biophysics and Bioengineering,* Vol. 6, F. M. Richards, "Areas, volumes, packing, and protein structure," pages 151–176. Copyright 1977 by Annual Reviews. www.annualreviews.org.

same polypeptides when denatured. Smaller conformational changes, e.g., the well-known T versus R states of hemoglobin, still can result in appreciable changes in exposed surface area. Also, the calculated area is strongly dependent on the choice of the probe radius and on the sphere sizes chosen to represent the atoms of the macromolecule. It is not legitimate to compare calculated ASAs from different molecular systems unless the same conventions are used for all probe and atomic radii. Figure 1.5, taken from Richards' 1977 review [33], shows how a change in probe radius from R_1 to R_2 will affect the apparent accessible surface. The figure also shows how certain parts of the macromolecular surface may not be in direct contact with solvent, because of crevices or dimples in the surface.

A useful factoid: the water surface density (water molecules per unit area) for proteins is about 0.11 to 0.15 $H_2O/Å^2$. This corresponds to an apparent cross-sectional area of a single water molecule of about 6 to 9 $Å^2$ on the protein surface.

1.3.2 CONVERGENCE OF FUNCTIONAL GROUPS

1.3.2.1 The Proximity or Chelate Effect

Polyvalency, the use of multiple combining groups to form a complex, is a potent way to increase both the strength and the specificity of binding. Early work on complexes of metal ions with various organic and inorganic ligands showed that the strength of association increased dramatically with increasing numbers of points of attachment of the ligand to the metal ion. For example, ethylenediamine tetraacetate (EDTA) is well known for its ability to form stable complexes with divalent metal ions. EDTA contains two amino and four acetate groups for a total of six possible points of contact with a metal ion; accordingly, it will form much stronger complexes with Mg^{2+} or Ca^{2+} than can individual acetate or amine molecules, which have only one or two points of contact. The increase in complex stability with greater number of combining groups in the ligand is known as the *chelate effect* [34,35].

In a chelate the individual interactions of the functional groups with the metal ion are rather weak, and so it is hard to explain the increase in complex stability strictly on energetic grounds. However, binding of the metal ion to one functional group on the chelator virtually guarantees further ionic interactions with other functional groups on the chelator. That is, the chelating agent brings together all these weak interactions into a very small spatial region, by virtue of its own internal covalent bonds. This might be considered to be a sort of entropic cooperativity that concentrates the functional groups for binding the metal ion, since the gain in free energy from forming one interaction point is augmented by the effectively higher local concentration of the other functional groups covalently linked to the chelating agent (see Figure 1.6). Viewing it differently, there is an entropic penalty in assembling many

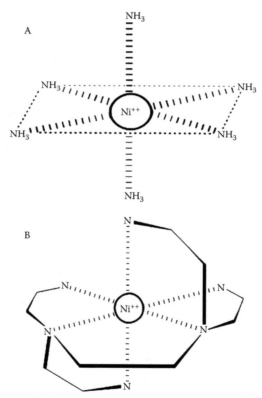

FIGURE 1.6 Chelation of Ni^{2+} by penten [tetrakis-(β-aminoethyl)-ethylenediamine] versus ligation with six molecules of ammonia. Here the chelate effect raises the affinity by more than 10 orders of magnitude: for addition of six ammonia ligands, $\log K = 8.49$, while $\log K = 19.30$ for ligation with penten (from *Helv. Chim. Acta* Vol. 35, G. Schwarzenbach, "Der Chelateffekt," pages 2344–2359 [1952]). A. Ni(II)-hexamino complex. B. Ni(II)-penten complex. (Hydrogens not shown in order to emphasize chelating interactions.)

individual small ligands around the central metal ion, due to losses of translational freedom of the many independent particles being grouped together. (There may also be losses in rotational and vibrational degrees of freedom for each of the molecules in the assemblage.) This penalty does not appear with a chelator, since the entropic penalty of gathering the groups together has been paid during the synthesis of the covalent bonds linking the combining groups.

An extension of the chelate effect, useful in thinking about binding specificity, is to consider further restrictions on arrangements of groups in the chelating agent so as to aim them in particular directions. This *preorganization* of a binding site is a common theme in achieving binding specificity with macromolecular receptors [36]. Binding sites on the macromolecule are already preorganized, with the entropic cost of their functional group orientation paid when the macromolecule was originally synthesized and folded.

Elegant studies by Cram on host-guest complexes have shown how differences in binding constant can be accounted for by the free energy demands of organizing the combining groups on the chelating agent [37]. For example, consider the two anisole hexamers, one linear and the other joined onto itself to form a spherical cavity (Figure 1.7).

The cyclic hexamer spherand forms a cavity that just fits around a desolvated Na$^+$ or Li$^+$ cation, so that the spherand enjoys the energetic gains from making contact with all six of its anisole moieties. The linear hexamer can indeed fold into such a ringlike structure; however, when one takes account of allowed bond rotations,

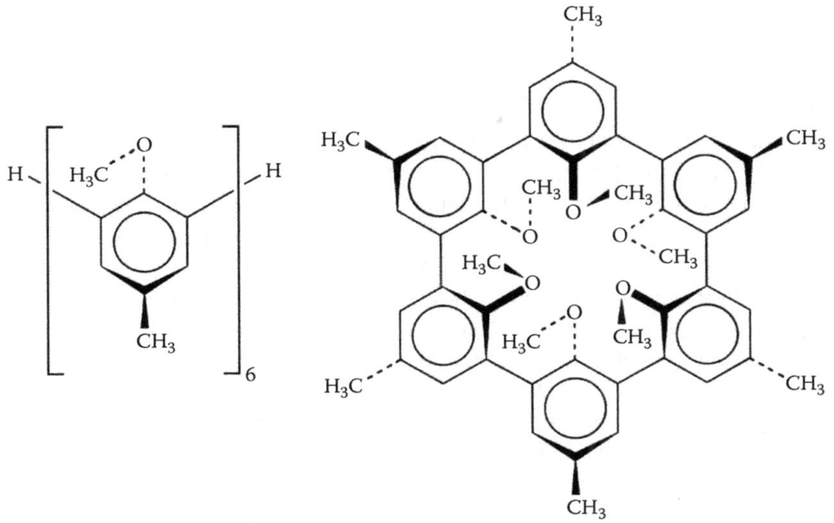

FIGURE 1.7 The cyclic anisole hexamer binds a single desolvated Na$^+$ or Li$^+$ cation much more strongly than does the linear hexamer. Adapted with permission from *Journal of the American Chemical Society* Vol. 107, D. J. Cram and G. M. Lein, "Host-guest complexation. 36. Spherand and lithium and sodium ion complexation rates and equilibria," pages 3657–3668. Copyright 1985 American Chemical Society.

among the 1024 possible conformations for this linear oligomer there are only two such ringlike arrangements. It is thus very unlikely that the oligomer would spontaneously fold up into one or the other of these two ring structures. It seems far more likely that, in binding the cation, the dominant linear oligomer conformations would have only two, three, or perhaps four of the anisole moieties contacting and wrapping around the cation; the remaining moieties in the oligomer would not be so restrained. As a result, the linear hexamer would not gain as much energy from binding the cation, and its affinity for the cation would be correspondingly less. Bringing the last few anisole moieties into contact (to gain a favorable enthalpy change from these contacts) would involve an entropic cost of restraining the entire linear oligomer in an improbable conformation, with an unfavorable contribution to the overall binding free energy.

Compared to the linear hexamer, the cyclic hexamer has greater preorganization in its binding site. All six anisoles will contribute energetically to the binding free energy. There is no entropic cost of bringing errant anisoles into contact with the cation; they are already held in position covalently. Thus, the cyclic hexamer binds Na^+ and Li^+ cations with high affinity; the binding constant for Li^+ in $CDCl_3$, saturated with D_2O, is greater than 7×10^{16} M^{-1}, corresponding to a favorable binding free energy change of more than 96 kJ/mol. However, in this solvent the linear hexamer binds Li^+ with a binding constant below 2.5×10^4 M, corresponding to a $\Delta G°$ less than -25 kJ/mol. As a result of the preorganization of the binding site into a rigid cyclic array of contacts, the binding is enhanced by -71 kJ/mol, an impressive effect.

Preorganization can artificially increase the local concentration of points of contact between ligand and receptor far beyond what one might naively expect. For example, the highest bulk molecular concentration achievable in aqueous solution is 55 M, the concentration of pure water itself. Concentrating the contacts to 55 M (relative to, say, a 1 M standard state) would increase the binding constant by a factor of 55; the corresponding change in the binding free energy is about -10 kJ/mol at 298 K. This 55-fold increase in concentration amounts to a loss in entropy on the order of 33 J/mol-K, which can be thought of as ΔS_{preorg}, an entropic cost of preorganization.

However, in the binding site the chemical groups are not free to move about, even over this small volume of solution. They are instead restricted by covalent bonding to particular spatial locations and orientations with respect to one another. This overall restriction on group arrangements can lead to much greater values of ΔS_{preorg}. For example, in the gas phase, the formation of a covalent bond between two molecules, a bond that severely constrains the mutual proximity and orientation of the two partners, may cause a loss in translational entropy of up to 120 J/mol-K [38]. This is much larger than the simple concentration effect just described. Similar results are obtained in the liquid phase. With the cyclic anisole hexamer-Li^+ system described above, if the enhancement of binding were due solely to preorganization, the entropic contribution would be -239 J/mol-K, a very substantial entropic effect indeed.

Many of the estimates of effective or local concentrations are based on a comparison of a bimolecular reaction (e.g., condensation of two acetates to form acetic anhydride) with a unimolecular reaction (the internal condensation of succinate to form succinic anhydride), using a ratio of the respective rate or equilibrium constants. This ratio will have the dimensions of a concentration, hence the use of the term *effective concentration*. Page and Jencks [38] present a classic illustration of the effects of preorganization in the ratio of the equilibrium constant for the formation of succinic anhydride compared to that for formation of acetic anhydride: the ratio is 3×10^5, implying that in succinate the effective concentration of carboxylic acid groups is 3×10^5 M, which, of course, cannot be physically real.

In extreme cases, in which the groups are rigidly aligned for maximal interaction and there is no strain on the conformation, the effective or local concentration of these groups can reach 10^8 M. This degree of concentration is completely nonphysical, since in aqueous solution no solute can be present in more than a few tens of molarity concentration. This exceptionally high effective concentration, compared to bulk solvent, is due to the large degree of disorganization in any solvent (along with the interstitial volume in the disorganized liquid), compared to the rigidly organized and closely spaced contacts in the binding site. Flexibility in the complex will reduce the effective concentration, but very substantial accelerations of organic and enzymatic reactions can still be explained on this basis.

1.3.2.2 Clefts as a Structural Motif for Binding Sites

Molecular clefts (depressions, concavities, etc.) are very often used as binding sites. A cleft provides a basis for size and shape selectivity: the ligand should match van der Waals surfaces with the receptor. Clefts provide a three-dimensional lattice of contacts, for alignment of H-bonds, ionic interactions, dipoles, and so on; that is, they provide the necessary three-dimensional complementarity of functional groups that generates both specificity and affinity [36].

There are several more reasons that clefts are desirable as binding sites. First, interfering solvent can be more readily excluded from a cleft than from a flat surface. Second, associated with solvent exclusion, there may also be a possible entropic gain, as highly organized waters leave such a site to become more disorganized in bulk solvent. Third, electrostatic forces can be focused by the difference in dielectric constant between the receptor and the solvent (Figure 1.8). These local electrostatic fields, which may help guide the ligand molecule into the proper position and orientation, might require appropriate distances between charged groups in order to generate the necessary field strength, direction, and range [39–41]. Finally, the walls of the site will tend to reduce diffusional escape by the ligand and to reflect the ligand back to the center of the site.

1.3.3 CONFORMATIONAL FLEXIBILITY

1.3.3.1 Microstates

Biopolymers such as DNA or proteins are small enough molecules that energy fluctuations start to become important in their properties and behavior [42,43]. Noncovalent interactions, with energies on the order of the thermal energy of agitation k_BT, are easy to break on an individual basis, and so they can be broken and remade very frequently. If one were able to observe and compare individual macromolecules at equilibrium in solution, it would be apparent that the solution contained a mixture of many different conformational states of the biopolymer (all equilibrating among

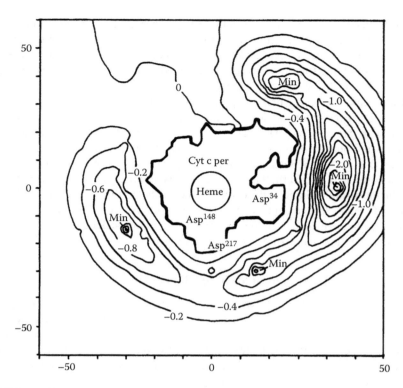

FIGURE 1.8 Charged groups around the folded surface of a receptor can focus electric fields and help guide ligands into the binding site. The calculated electrostatic potential (in units of $k_B T$) in a cross-section through cytochrome c peroxidase shows an electrostatic "channel" near the three principal docking sites for cytochrome c. Reprinted with permission from *Science* Vol. 241, S. H. Northrup, J. O. Boles, and J. C. L. Reynolds, "Brownian dynamics of cytochrome c and cytochrome c peroxidase association," pages 67–70. Copyright 1988 AAAS.

themselves). Also, for each of these conformations there would be a mixture of many different states of interaction with the ligand (again, all in equilibrium among themselves, and with the other biopolymer conformations, too). Since there is usually only a small energetic barrier to breaking or re-forming the weak interactions individually, there is a ready and rapid exchange of one binding state for another. In other words, the system fluctuates over multiple states, and any macroscopic binding measurement is an average, taken over the entire distribution of binding states, all of which differ just a bit in (free) energy.

These slightly different states of the system are called *microstates* (Figure 1.9). At equilibrium there will be a Boltzmann distribution of the molecules over these microstates. The interconversion among microstates should be relatively rapid, since there are only small potential energy barriers between them. Furthermore, large but relatively infrequent excursions in energy and volume (i.e., in conformation) are possible and can be explained by the rare coherence of a number of smaller fluctuations [42].

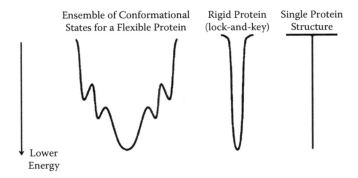

FIGURE 1.9 Microstates for a biopolymer. Flexible structures will have an ensemble of related conformations, separated by low-energy barriers (panel A), while more rigid structures will have a single-energy minimum (panel B). In the extreme case of a completely rigid structure, the distribution of energy states collapses to a delta function, symbolized by the line in panel C. Adapted with permission from *Molecular Pharmacology* Vol. 57, H. A. Carlson and J. A. McCammon, "Accommodating protein flexibility in computational drug design," pages 213–218. Copyright 2000 ASPET.

There is now considerable evidence to support the notion of an equilibrium distribution of related conformations or microstates for a macromolecule and its binding site(s), rather than a single conformation alone [44–46]. This includes the crystallization of certain proteins with two different conformations of the same protein in a single unit cell (e.g., the enzyme rennin [47]); bulk kinetic measurements on enzyme-catalyzed reactions, consistent with the existence of multiple intermediates and conformations [45]; NMR relaxation studies that detect conformational fluctuations in the active site on a timescale that correlates with substrate turnover (e.g., cyclophilin A [48]); fluorescence lifetime studies on receptors that detect oscillations around a single detectable conformation, with a change in conformation upon binding agonist but not antagonist (e.g., the β_2 adrenergic receptor [49]); existence of constitutively active receptors and partial agonists and inverse agonists for drug receptors, especially those coupled to G-proteins [50,51]; and single molecule studies on enzymes and other proteins that show spontaneous switching over two or more conformations [52].

Fluctuation over related conformations permits structural flexibility in the biopolymer, without loss of essential structural features. Thus, the overall biopolymer structure can be maintained while allowing for dynamic transitions in local conformation, which may be important for the proper functioning of the biopolymer. Also, fluctuations can provide a small molecule with rapid but transient access to regions in the interior of the macromolecule, on the millisecond to microsecond timescale. This has, in fact, been proposed as the mechanism by which carbon monoxide or oxygen molecules find the heme group in myoglobin [53]. Fluctuations in how local structures are hydrated may also permit the macromolecule a greater range of conformations, for more (or perhaps less) discrimination in binding. Figure 1.10 shows some other possible effects of ligand binding on the microstate distribution of a population of macromolecules.

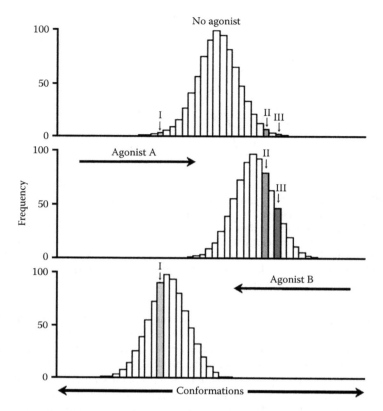

FIGURE 1.10 Different ligands may select and stabilize single conformations from the available spectrum of conformations of the receptor (panel A). By stabilizing a particular conformation, a ligand may shift the spectrum of equilibrium receptor conformations (panel B), and different ligands may shift the distribution in opposite ways (panel C). Adapted from *Trends in Pharmacological Science* Vol. 18, T. Kenakin, "Agonist-specific receptor conformations," pages 416–417, copyright 1997, with permission from Elsevier.

If the binding region is flexible (that is, has available to it a set of microstates for different, but related, conformations), then binding will not require disruption of a stable structure. This will minimize the activation energy for binding. It may also allow the site to accommodate a variety of different ligands with much the same affinity. There are many instances known of such systems [44,46]. A well-known example is the protease trypsin; this enzyme cleaves the peptide bond after amino acids with side chains carrying a positive charge (Lys and Arg), and can act on a wide range of proteins with (presumably) different shapes of the substrate [54]. Another instance is found with multidrug-efflux transporters, which show considerable breadth in specificity as they act to expel toxic compounds with widely diverse chemical structures [55]. As a third instance, the structural plasticity of the recognition domains of antibodies and T-cell receptors provide the immune system with much of its flexibility in recognizing diverse antigens [56].

If, however, binding results in the ordering of a (flexible, conformationally fluc-
tuating) region of the binding site, this will create an entropic debit. Some of the
binding energy must be dissipated to cover this entropic cost (consider the example
above of the cyclic and linear anisole oligomers binding cations). This means that the
ligand-receptor complex in these cases will be less stable (in terms of free energy)
than would be suggested by a simple sum of the favorable energetic interactions in
the complex. Fluctuations provide an alternative mechanism to a strictly mechanical
linkage of a series of small conformational changes in the receptor. Certainly, there
are systems in which the mechanical view is appropriate. Such systems, however,
typically do not involve qualitative changes in the structural ordering of the receptor, and one can trace more-or-less easily the propagation of conformational changes
through the structure. But there are others in which the mechanical linkage is not
at all clear, in which, for example, a disordered peptide chain folds down to bury a
ligand or substrate molecule (e.g., the *Eco*RV endonuclease wrapping around DNA
[57], or streptavidin enveloping a molecule of biotin [58]). These conformational
changes may also be quite difficult to predict in advance, related as they are to principles of protein folding and of ligand conformational changes in solution and in the
medium of a binding site.

1.3.3.2   Hydration and Flexibility

The hydration and the motion of biopolymers go hand in hand. It is well known
that, upon full dehydration, enzyme activity is often reduced or eliminated. Conversely, addition of a small amount of water, just enough to provide a single layer
of hydration, is often enough to restore full activity in enzymes [59]. Single water
molecules might act as individual lubricating elements between parts of a biopolymer, maintaining van der Waals and hydrogen-bonding contacts and so permitting
local motions of the biopolymer to occur without any substantial energetic penalty.
Furthermore, the freedom of motion of the hydration layer, with a variety of nearly
isoenergetic states, may allow small fluctuations or even fairly gross movements in
the protein, and so contribute to the continued action of the enzyme. In general, there
will be coupling of motions of a biopolymer and the motion of its waters of hydration. This may include cooperative motions of entire clusters of water molecules.
These cooperative motions would naturally be slower than those of individual water
molecules, but they might be important in coupling hydration to the motion of large
domains of the biopolymer, since there would be at least a rough match of the time-
scales involved. In this way the surface layer of water may act as a lubricant, to ease
large conformational changes in a hydrated biopolymer.

1.3.3.3   Time and Distance Scales

Since the flexibility of macromolecules is often important for their proper biochemical functioning, how large are the motions here, and how fast do they occur? In
proteins the timescale of local motions extends from about 5×10^{-12} sec for methyl
group rotation, through vibrational motions of the α-carbon backbone (correlation
times around 10^{-10} sec), to isomerization of proline residues, which can require up
to 10^{-2} sec. Whole-body rotation of a typical protein occurs with correlation times

on the order of 10^{-6} sec. Large-scale unfolding (denaturation) of the protein can be either very rapid (milliseconds or faster, especially for smaller proteins) or quite slow, with some proteins showing relaxation times on the order of hundreds of seconds [60–64].

Conformational changes in protein-ligand binding can involve considerable spatial displacements. For example, the closing of the two domains of citrate synthase around a substrate molecule involves an 18° rotation of one domain with respect to the other, and shifts of up to 15 Å for some atoms [65]. Binding of the substrate analog N-phosphonacetyl-L-aspartate to the enzyme aspartate transcarbamoylase induces a 12 Å expansion of the enzyme and causes rotations of about 15° of some subunits [66]. Many other proteins (e.g., antibodies, contractile proteins, etc.) can show similarly large movements of domains [63,64]. These proteins often accomplish their functions, whether catalysis or simple binding, on the microsecond timescale, implying that the large-scale motions must likewise occur at least that rapidly.

If energetic factors dominated the binding process, it could be that ligand binding would tend to induce a local tightening of receptor structure around it, in order to maximize energetically favorable contacts. However, at least in some cases it appears that binding may instead induce large-scale, low-frequency delocalized motions (so-called soft modes [67]), leading to a favorable increase in the entropy. In fact, the notion that translational and rotational degrees of freedom may be converted during binding into soft modes of vibration, has long been discussed [38,68–74]. Also, a small change in protein structure, as might be caused by a replacement of a single amino acid, can strongly affect the internal vibrations of the macromolecule [75].

As with proteins, so with nucleic acids: the local motions (vibrations, rotations) of individual atoms and small chemical groups occur on the picosecond timescale. Motions on a larger scale, for example, bending and twisting motions of the DNA double helix or opening of the base pairs, can have a broad distribution of relaxation times compared to local motions, since these motions involve more atomic units and more chemical bonds. NMR experiments show local angular motions of the bases in a B form DNA double helix to lie in the range 8 to 10° [76]; these motions take place on a timescale faster than 10^8 s^{-1}. Measurements of the torsional flexibility indicate fluctuations in the winding angle between adjacent bases of 3 to 5 Å, with motions on the nanosecond timescale [77].

Fluctuational opening of the base pairs results in an open fraction at room temperature or slightly above of about 10^{-5} (for base pairs in the helix interior; pairs at the ends of the helix open much more frequently, by perhaps a factor of 10^4). The lifetime of an open base pair (again, in the interior of the helix) is about 10^{-6} sec, so each base pair must open transiently about 10 times per second on the average [78].

Nucleic acids can also undergo large conformational changes upon binding a ligand. The binding of intercalating agents such as acridine or ethidium seriously distorts the DNA double helix [79]: the intercalator pushes the base pairs apart (normal base pair separation is 3.4 Å, but this increases to about 6.8 Å around the intercalator), and it unwinds the helix (in the case of ethidium, by about 26° for each molecule bound). This is accompanied by changes in the conformation of the phosphodiester backbone and in the torsion angle of the glycosidic linkage at the 3′ end. Some complexes may also involve conversion of Watson-Crick pairing to Hoogsteen pairing, as found with the *bis*-intercalator triostin A [80]. The insertion reaction itself occurs in the millisecond time range for simple compounds with uncomplicated binding mechanisms, such as proflavin [81]. Other intercalating agents, e.g., daunomycin [82] or nogalamycin [83], may follow slower and more complex mechanisms for binding, which involve conformational changes in the drug or more extensive conformational changes in the DNA than is required for the proflavin intercalation.

DNA interactions with large proteins can result in quite large amounts of bending or other distortions of the DNA. Certain gene-regulatory proteins can induce DNA curves approaching a 90° bend or more; e.g., the CAP protein of *E. coli* [84], the Cro protein of phage λ [85], or the TATA binding protein [86]. The wrapping of DNA around histones to form nucleosomes involves bending the DNA around a protein core of about 110 Å radius, for 1¾ turns [87]. The "looping" of DNA, in which a single protein or set of proteins may attach simultaneously to two separated sites on the DNA, is now recognized as a major factor in gene regulation, site-specific recombination, and DNA replication [88–90]. Looping can involve tens to thousands of base pairs, and may require both torsional and lateral flexing of the DNA chain.

Enzymes that act on nucleic acids often induce local distortion of the nucleic acid conformation. Of course, DNA and RNA polymerases must denature and unwind the double helix over an extended region, at least transiently, in order to carry out their respective functions [91,92]. Other enzymes produce notable local disturbances of the nucleic acid. For example, the restriction endonuclease *Eco*RI recognizes the "canonical" DNA sequence GAATTC, cleaving the DNA in each strand between the G and A residues. The tightly bound protein induces notable localized perturbations of the DNA backbone within the GAATTC sequence, with overall unwinding of the DNA helix by about 25 Å and introduction of a bend of about 23 Å in the axis of the helix [93,94]. As another example, the *Hha*I DNA methyltransferase and the *Hae*III methyltransferase methylate cytosine residues by forming a complex where the unmethylated base is extruded from the helix (with local unpairing of the bases), and inserted into a site on the enzyme where it receives a methyl group from the cofactor *S*-adenosyl-*L*-methionine [95,96].

1.4 BINDING SITES ON PROTEINS

1.4.1 MACROMOLECULAR STRUCTURES AND THE PROTEIN DATA BANK

The Protein Data Bank (PDB) was established in 1971 at Brookhaven National Laboratory as a repository for crystal structures of biopolymers [28]. Initially a resource for specialists, with only a handful of structures available, it is now used by a wide group of scientists and students, at all levels of expertise, thanks to access through the World Wide Web and free software for visualizing, manipulating, and analyzing the structures in the database. Currently the repository is managed by Rutgers University, the Supercomputer Center of the University of California, San Diego, and the National Institute of Standards and Technology. In addition to crystal structures of biological macromolecules, the PDB also contains structures determined by NMR [30]. A similar repository was set up in 1991 for structures of nucleic acids (the Nucleic Acid Database or NDB [31,32]). Both contain numerous examples of macromolecule-ligand complexes. Much of what is known about the characteristics of binding sites on proteins and nucleic acids comes from analyzing data stored in these two databases. In addition, there are some more specialized databases, especially for ligand-macromolecule complexes: Relibase [97] has structures of complexes but no information on binding affinities; the Binding Database [98] has binding affinity data on a variety of binding systems; and the Ligand-Protein Database [99], the Protein-Ligand Database [100], PDBLIG [101], and the PDBbind Database [102] all collect and correlate macromolecule-ligand complex structures and binding affinities.

1.4.2 SMALL MOLECULE SITES

Viewed broadly, globular proteins are folded such that nonpolar side chains are buried away from contact with solvent, while polar side chains tend to be found on the surface, exposed to solvent [6,103–105]. Binding sites in clefts and pockets are exposed to solvent, so one might expect to see a preponderance of polar residues lining the walls of such sites. On the other hand, a notable feature of many drug molecules is their hydrophobicity, which would indicate that their binding sites might be largely formed from nonpolar moieties. So it is reasonable to ask, Is there any pattern or bias in the distribution of amino acids around and in binding sites? In a survey of 50 protein-ligand complexes (including both large and small molecule ligands), Villar and Kauvar [106] found that Arg, His, Trp, and Tyr tend to be substantially overrepresented at binding sites, compared to their occurrence frequency in proteins generally. In a more extensive study, Bartlett et al. [107] analyzed the crystal structures of 178 catalytic sites of enzymes for amino acid frequency. His, Asp, Arg, Glu, Lys, and Cys (all having polar side chains) were found more frequently in the sites of these enzymes than would be expected, based on overall amino acid frequency in the proteins. Histidine was especially prominent, constituting 18% of all catalytic residues. Also, certain residues were apparently disfavored for active sites: Phe, Leu, Met, Ala, Ile, Pro, and Val; all are notable for their nonpolar character. Some amino acids were apparently neutral with respect to distribution: Asn, Gly, Ser, Thr, Gln, and Trp. The neutrality of Ser and Thr is surprising in view of their role in covalent catalysis mechanisms.

Regarding conformational states of the active-site residues, Bartlett et al. found that the "coil" conformational state was the most common. In contrast, for the average amino acid in this set of proteins, the alpha helix was the most common conformation. Presumably the coil conformation allows flexibility in accepting a substrate, adjusting to form the transition state, accommodating extra molecules (e.g., water), and releasing the product. According to the crystal structures, however, these residues typically held position more rigidly than the average, even though they were in a (more-or-less disordered) coil conformation; thus, alignment is still important for binding and catalysis. Finally, the majority of catalytic residues were in a cleft, with limited exposure to solvent. Desolvation of the molecular surfaces at the interface can thus be expected to frequently play an important role in the binding process.

How large are these sites, and how much solvent-exposed area is buried in the complex? The active sites of enzymes are typically found in the largest clefts on the enzyme's surface [108]. Janin and Chothia [109] computed the loss of surface area upon binding of coenzyme (flavin mononucleotide or nicotinamide adenine dinucleo-tide) in three different enzymes. Upon binding, the coenzymes bury about 600 Å^2, with a comparable area on enzyme also removed from exposure to solvent. This agrees with the finding of Pettit and Bowie [110] that small functional sites on pro-teins with an area less than about 600 Å^2 are notably rougher than larger sites. These rough sites (e.g., pockets or clefts) would be preferred for binding of small ligands. Docking a small ligand at a flat surface would involve contacts only along one side of the ligand, and this might not be enough surface area to stabilize a complex by van der Waals interactions and solvent release alone. It also might not offer enough speci-ficity or selectivity in binding, since only one side of the ligand would be involved. A cleft would allow more surface-to-surface contacts, and so help to stabilize the com-plex; a cleft or pocket would also make contact over more of the three-dimensional surface of the ligand, and so would lend more specificity to the binding.

In developing a predictive scheme for noncovalent binding sites for small druglike molecules on proteins, Hajduk et al. [111] selected 57 potential small-molecule binding sites from a group of 23 proteins. These sites included 28 known or NMR-identified ligand-binding sites, so-called "positive pockets," and 29 additional sites for which no binding was observed by NMR, so-called "negative pockets." With software, Hajduk et al. extracted geometric and chemical features of these pockets, and correlated these with the "positive" or "negative" character of the pocket; they then used these statisti-cal correlations to attempt to predict the occurrence and location of binding sites on a set of proteins not used in developing the original correlation model. They found that the negative pockets had a significantly smaller volume and less nonpolar surface area; these pockets were less rough, and tended to be longer and narrower than the positive pockets. Interestingly, for the binding of druglike molecules it appears that polar residues on the protein play a minimal role; pocket hydrophobicity and pocket shape dominate their model for predicting positive pockets. Hajduk et al. concluded that "endogenous ligand-binding sites tend to be the largest, most hydrophobic, and most geometrically complex pockets on the protein surface" [111].

1.4.3 PROTEIN-PROTEIN INTERFACES

Protein-protein interfaces differ qualitatively from protein–small molecule binding sites. First, these interfaces involve fairly substantial amounts of surface area; the aver-age interface typically covers 1600 (±400) Å^2, with roughly equal contributions from each partner [112]. This is much larger than the average surface area of a binding site for small molecules (typically about 600 Å^2 or less). Second, the average protein-pro-tein interface is notably flatter and less rough than the typical small molecule binding site [110]. The size and the roughness of these interfaces have been of considerable interest, especially to those chemists trying to design drugs that would interrupt pro-tein-protein association. Small molecule binding sites tend to be in clefts or pockets, so if there are no such regions within a protein-protein interface, then it may not be feasible to design a compound to bind there that would directly block the protein-protein association. Instead, the rational drug designer would have to seek rougher sites outside the interface for binding, hoping that occupancy of one or more of these peripheral sites would break up the association.

Some other characteristics of protein-protein interfaces are that they generally have few "gaps" or vacancies, so that the juxtaposed surfaces generally match each other sterically [113]. The boundary of the interface is quite variable in shape; on the average it is only roughly circular, with a ratio of principal axes around 0.7, but some complexes have quite elongated interfaces, with axial ratios down to 0.25. The secondary structure of these interfaces has approximately equal contributions from helical, strand, and coil residues. Packing of interfacial residues is comparable to that in the interior of the protein [112], with perhaps greater structural rigidity for conserved residues that form local surface asperities or pockets that are geometrically complemented by residues of the partner protein [114].

Enzyme-inhibitor complexes and permanent heterocomplexes tend to have the most complementary surfaces with the least amount of gap, and homodimeric interfaces are more planar, and tend to have more hydrophobic residues, than heterodimeric interfaces [113]. (Here, "permanent" means a complex whose subunits do not dissociate appreciably in functioning, while "transient" as used below means that the proteins regularly associate and dissociate as they function.) Homodimeric protein subunits are rarely found as free monomers and they rarely function as monomers, so interfacial burial of a pair of hydrophobic patches, making close contacts, would serve them well. On the other hand, the subunits of heterodimeric complexes frequently function as free monomers, so there would be an energetic disincentive for them to carry a large hydrophobic patch that would be exposed in the monomeric state.

Ofran and Rost [115] applied sophisticated information-theory-based methods to sort structural data on contacts between residues into six categories, based on structural or functional association. These categories were, respectively: (1) intra-domain interfaces, or contacts between residues in the same structural domain; (2) interfaces between different domains within one chain; (3) interfaces between identical chains in a permanent complex; (4) interfaces between identical chains in a transient complex; (5) interfaces between nonidentical chains in a permanent complex; and (6) interfaces between nonidentical chains in a transient complex. The six types of interface were distinct in their respective amino acid compositions, but some types were more distinctive than others; the most distinctive interface type was a transient interface between identical protein chains. All interface types differed significantly in composition from the average amino acid frequency found in the SWISS-Prot protein sequence database [116], and they differed as well from the average composition of solvent-exposed surfaces of proteins. Most of the amino acids with large hydrophobic side chains were favored at interfaces over the average. Conversely, Ser, Ala, and Gly were under-represented, compared to the average. As for charged residues, Lys was generally under-represented at interfaces compared to average protein composition, while Arg was generally over-represented. Ofran and Rost found that Trp was under-represented in one class of interface (between transiently interacting identical proteins), but over-represented in the other five types of contacts.

1.4.4 BINDING 'HOT SPOTS'

Certain regions of protein-protein interfaces or ligand-receptor interfaces often contribute much more than would be expected to the binding energy; these regions, or the residues in them, are referred to as binding "hot spots." They can be identified by systematic chemical modification or by mutational studies. The general procedure is to alter a protein residue or set of residues by chemical reaction or by mutation, then to determine the binding affinity to see how much the affinity has changed. A hot spot can be defined as "a residue that, when mutated to alanine, gives rise to a distinct drop in the binding constant (typically tenfold or higher)" [117]. With the apparatus of modern genetic engineering and structural biology, especially with the wealth of structures available from the PDB, it is now possible to generate and compare moderately large sets of mutationally related proteins. This has been done for a wide variety of proteins (enzymes, hormones, protein inhibitors of enzymes, drug receptors, antibodies, etc.), and the results have proved quite interesting. Currently, the number of reports concerning direct chemical modification of sites has been eclipsed by those related to the mutational scanning methods, so this summary will concentrate on results from the scanning method.

There are two main divisions of the mutational method: homolog-scanning mutagenesis and alanine-scanning mutagenesis [118,119]. In the first, entire segments of the protein chain are replaced by sequences from a homolog of the receptor that has reduced affinity for the ligand, which will likely entail multiple simultaneous changes in the (mutated) protein sequence. Presumably, this homologous replacement would not have much effect on the folding or stability of the protein overall, but there would be some effect on the affinity for ligand through replacement of some residues, which could then be used to identify the key residues for binding. In the second, amino acids in the region of interest are sequentially replaced by alanine, to generate a set of proteins with single mutations in their sequence and, presumably, with differences in their binding. Replacement by alanine does not alter the main-chain conformation, but it does eliminate the side chain beyond the β carbon; thus, this method should probe the role played in binding by more distant side chain atoms. No new ionic charges are introduced and no bulky groups are inserted; thus, the electrostatic and steric effects of the replacement should not be extreme and protein folding and stability should be much the same for these single mutants as for the natural wild-type protein. This method also avoids the complications of possible multiple simultaneous mutations that arise with the homolog-scanning method.

The interface often has a central region that contains a number of hot spots [120,121]. Furthermore, the hot spots are often dispersed within this central region, rather than clustered into a single compact site. These hot spot residues are often hydrophobic, and they tend to make contact across the interface with hot spots on the other side [120]. The fit across the interface is rarely perfect, and there may be unfilled pockets present; these may account for 5 to 20% of the area of the interface [114]. The imperfect docking due to unfilled pockets may allow for flexibility and adaptability in the protein-protein interface. Typically, the amino acids forming the unfilled pockets are not highly conserved. On the other hand, residues that interdigitate tightly across the interface tend to be conserved, and the residues forming these complemented pockets tend to be less hydrophilic than those in the uncomplemented pockets [114].

It is not a surprise that hot spots tend to be structurally conserved residues at the interface. Conserved polar residues tend to lend rigidity to the complex, which might help to reduce entropic penalties associated with looser complexes. Surrounding residues tend to be more flexible, and may serve to block solvent from the hot spot [120]. As for the different amino acids identified as hot spot residues, Ma et al. [121] found that Trp leads the list, followed by Arg, Tyr, Leu + Ile (the pair considered a single moiety), Asp, His, Pro, and Lys (see the results of Ofran and Rost [115], summarized above). The conservation of a Trp residue across a protein family probably indicates its participation in binding; Trp is large, nonpolar, and aromatic, and would provide a distinctive recognition feature to mark a binding site. Interestingly, Met also tends to "prefer" binding sites over the general protein surface. This might stem from the ability of Met to form weak hydrogen bonds, using its sulfur atom. The sulfur atom is also more polarizable than the average atom in a binding site, which could be important for electrostatic interactions.

While hot spot residues appear to dominate the binding energetically, they are perhaps not so important for binding specificity. It appears that the other regions of the binding site, which don't contribute so much to the energetics, serve more to direct the binding specificity. These residues may form a sort of "O-ring" around the central hot spot region, with the O-ring sealing off access by (polar) solvent to the central (largely nonpolar) region [120]. This occlusion of solvent could help to increase the strength of hydrogen bonds and electrostatic interactions in the central region, since the dielectric constant there would now be considerably lower than if solvent were allowed to penetrate. With stronger interactions, the dissociation of the protein partners would be slowed.

Protein-protein associations generally occur with rather high bimolecular rate constants, on the order of 10^6 $M^{-1}sec^{-1}$; this is, however, much slower than the rate constant (about 7×10^9 $M^{-1}sec^{-1}$) for diffusion-limited association of uniformly reactive spheres of the size of a protein with a radius of 18 Å. Proper alignment of the protein interfaces is obviously required for a stable association, and so the association rate should be slower than this diffusion-controlled limit. However, relying on random encounters to align the interfaces of two proteins for proper docking (within 2 Å of the correct position; displacements greater than this would likely not be energetically stable) would lead to a rate constant of only 7×10^2 $M^{-1}sec^{-1}$, which is slower than the experimental values by at least a factor of 10^3 [122]. The explanation for the high rate of association, which is still not diffusion-limited, is that after a three-dimensional random walk leading to a protein-protein collision, the proteins are trapped by cages of solvent molecules so that they lie in close proximity to each other, where they undergo multiple collisions with each other. These nonspecific associations, termed *encounter complexes,* persist for several nanoseconds and allow the two molecules to reorient themselves by rotational diffusion, until proper alignment is achieved [122]. Specific attachments or interactions would only come into play when the two proteins were well aligned.

This explains why many proteins have roughly the same bimolecular rate constant for protein-protein association. The rate of association seems to be set primarily by diffusional considerations and not by specific interactions. Since complexes obviously have different overall binding affinities, then affinity is determined primarily by the kinetics of dissociation, not association. Wells [123] suggests that it is here that binding hot spots are important: these are specific interaction points whose strengths determine the rate of complex dissociation, not association.

1.4.5 PROTEIN SURFACES THAT BIND DNA

As more structural information has been compiled, it has become apparent that protein-DNA interfaces have their own set of characteristic features, distinct from protein-protein interfaces and small molecule binding sites. Jones et al. [124] surveyed 26 protein-DNA complexes and 21 B-DNA structures, taken from the PDB and NDB; this was followed in 2000 and 2001 by two much more extensive reviews [125,126]. The proteins surveyed here were transcription factors and enzymes that act on DNA. Jones et al. were able to differentiate three general structural classes for the proteins, but Luscombe et al. [125,126] found eight different structural classes, each with a distinct structural feature or motif involved in DNA site recognition and binding. The eight groups so distinguished were (1) HTH, or helix-turn-helix; (2) zinc-coordinating, including so-called zinc fingers; (3) zipper-type; (4) other α-helix; (5) β-sheet; (6) β-hairpin/ribbon; (7) "other"; and (8) enzyme. These eight groups could be further classified into 54 families, based on structural homology comparisons [125].

The area of the binding sites on these proteins is substantial, with the amount of buried ASA ranging from 618 to 2800 $Å^2$ [124]. These regions are substantially more polar and hydrophilic than the typical protein-protein interface. There are many hydrogen bonds per unit surface area (between 0.9 and 2.4 hydrogen bonds per 100 $Å^2$ of interface ASA), and many buried water molecules that form bridges between the protein and DNA (up to 1.94 buried water molecules per 100 $Å^2$ of interface ASA). The protein side of the interface is formed from discontinuous segments of protein chain, with sites having between 2 and 16 such segments. The degree of segmentation here is notably higher than for the average protein-protein interface. All types of protein secondary structures are represented at the interface, with the proteins using alpha helices, beta sheets, and loops to make contacts with the DNA.

As for amino acid composition of the binding sites, Jones et al. [124] found that, compared to the frequency of occurrence on the protein surface overall, the most highly preferred amino acids in binding sites are Arg and Thr, followed by Asn, Gln, Ser, Lys, and Gly. The later review by Luscombe et al. [126] also found Arg to be the most highly preferred, but classified Lys as the next most preferred, followed by Ser, Thr, Asn, Gln, and Gly. Some amino acids are strongly disfavored in the binding sites, notably Asp and Glu [124,126]. The negative charge carried by the side chain on Asp and Glu would, of course, not be favorable in close proximity to the sugar-phosphate backbone of DNA. While Trp is often favored in protein-protein contact hot spots, here it is strongly disfavored. As noted above, the sites are notably more polar than the typical protein-protein interface, and nonpolar and bulky residues (e.g., Leu, Ile, and especially Trp, but also Pro) are disfavored.

Binding most commonly occurs by contacts in the major groove; some proteins make contact in both grooves and a few make contact only in the minor groove. The DNA is often distorted in base-protein contacts. A common feature is a "kink," where one base step has a large roll; some complexes show two or more kinks. This may result in bending of the DNA helix (as with the cAMP receptor protein of *E. coli*, with a net bend of the helix of 90° [84]) or the bends may cancel out (as in the case of the *Eco*RI restriction endonuclease [93]). Bending of the DNA may be toward the major groove, compressing it and widening the minor groove, or toward the minor groove, compressing it but widening the major groove. Less common is a grosser distortion, e.g., the flipping-out or eversion of a base or base pair (as with the *Hha*I DNA methyltransferase [96]). In some cases, the protein changes conformation substantially while the DNA not much at all; an example here is the *Bam*HI restriction endonuclease [127]. In other cases both the protein and the DNA undergo substantial conformational changes (as with the γδ-resolvase [128]).

1.5 BINDING SITES ON NUCLEIC ACIDS

Nucleic acid binding sites offer three significant structural features that are important for their recognition by proteins and small molecules. First, there is the general shape of the nucleic acid molecule: helical or single-stranded, right- or left-handed helix, A form helix versus B form helix, major versus minor groove, etc. Second, there is the polyanionic sugar-phosphate backbone, with its high density of negative charge. And third, there is the sequence of bases in the site, with variations in the arrangement of possible contact points for hydrogen bonding, stacking interactions, etc.

The size of these binding sites has quite a wide range, and depends strongly on the nature of the ligand being attached. Small organic molecules will typically make contact with two to four bases or base pairs, though some specially synthesized ligands may be designed to make contact with double or triple this number of bases or base pairs. Gene-regulatory proteins generally use sequences of 15 or more base pairs. Then again, type II restriction endonucleases use short sequences, typically four to six base pairs. The difference in length here reflects differences in function. An operator sequence, for example, should be relatively rare, with perhaps a handful of such sites per cell, and a longer base sequence makes for a lower frequency of occurrence [129]. Conversely, restriction sites might be expected to be found rather frequently for more efficient protection against invading foreign DNA, and a site of four to six base pairs, with its high frequency of occurrence, will offer many opportunities for the endonuclease to cleave the foreign DNA, while not placing an undue burden on the host cell for methylating and protecting its own DNA.

Recognition sites for proteins are often repetitive and symmetric in sequence (when the sequence is inverted and repeated this is referred to as a *palindrome*). The proteins involved are often dimeric or tetrameric, and one subunit will typically recognize only a portion of the entire DNA site. In this way the cell avoids making a large protein to span the entire sequence, while generating a long enough DNA sequence to make it unique. Repetition of a sequence may also be important for "in-phase bending" to create a characteristic, recognizable shape, or to increase DNA looping probability, or perhaps to permit generation of a cruciform structure (see below).

RNA is known for its structural plasticity: in addition to the double-stranded A form helix, RNA can form triple helices and have single-stranded loops and bulges in the middle or at the ends of helices [130–132]. The folding of transfer

RNA molecules into their characteristic "L" shape is well known [130]; recognition of specific tRNA molecules by their respective aminoacyl synthetases involves base-specific contacts within the anticodon loop. Mismatched base pairs are a common feature in naturally occurring RNA, as is the presence of one or more of a variety of modified bases, including pseudo-uracil, methylated guanine and adenine, dihydro-uracil, and inosine.

DNA is also quite capable of taking on a variety of secondary and tertiary structures. In addition to the standard right-handed B form helix, under suitable conditions DNA can also form right-handed A helices, as well as the left-handed Z helix. Additionally, DNA is capable of forming unusual structures, including supercoiled DNA, triple and even quadruple helices. As with RNA, DNA molecules may contain single-stranded loops and mismatched bases or longer "bulge" regions [133–136]. One very striking configuration is the cruciform structure that is formed from palindromic sequences. Though less stable thermodynamically than the regular double helix, cruciform structures can be extruded by supercoiling the DNA first, and using the free energy stored in the supercoils to drive formation of the cruciform.

The same general principles should apply to site recognition in RNA as in DNA. Relatively little is known, however, about specific binding to double-stranded RNA (or to Z form DNA, for that matter), by comparison to the wealth of studies on B form DNA. As for single-stranded nucleic acids, binding of proteins to single-stranded nucleic acids typically involves pockets on the protein for binding specificity; see the preceding section for general features of such protein binding sites. This discussion will be limited to recognition features of helical double-stranded B form DNA.

Overall, the intact double-stranded B form helix is moderately rigid with respect to both torsional and bending motions [137–140]. However, it is quite possible for there to be rather sharp bends, even kinks, which arise spontaneously in the helix [141,142]. Introduction of single-stranded regions ("bubbles") or some local conformation with much greater flexibility would dramatically increase the bending and twisting flexibility of either DNA or RNA. Recent experiments support this concept [141,142], though it is not yet clear precisely what the local conformational change is (for example, this might involve unpairing a base and rotating it out of the interior of the double helix).

Such bends could, for example, be quite important in forming short or extended loops of DNA. By juxtaposing widely separated regions of the DNA chain, loop formation offers several possibilities for augmenting binding affinity and specificity of protein–nucleic acid interactions. For example, it allows room for auxiliary regulatory proteins to attach to a specific protein-DNA complex so that they might modulate the activity of the chief regulatory protein, as in DNA replication complexes with their multiple proteins and subunits. In bringing two specific DNA sites into close proximity to one another, looping can increase the chances that both sites will be occupied by a single protein or protein complex (see Section 1.3.2, on the chelate effect). Thus, a cell might be able to use lower protein concentrations to saturate key regulatory sites, while allowing the individual DNA sites to have rather modest association constants for the protein. This in turn would allow partial or transient opening, but facile reclosing of such complexes, so as to permit other cellular processes to proceed (e.g., DNA replication or recombination).

1.5.1 NUCLEIC ACIDS AS POLYANIONS: SALT EFFECTS IN LIGAND BINDING

The high charge density and linearity of the array of charges on nucleic acids distinguish them from most other biopolymers and in particular from proteins (some ionic polysaccharides, such as heparin, are linear and may carry a substantial charge, however). The negative charge on the phosphates creates a strong local electrical field that will attract counterions and repel coions. This field drops off rapidly, however, in physiological salt solutions. Still, there are electrostatic focusing effects in the grooves; the minor groove is especially attractive to positively charged moieties, and counterions will tend to gather in those regions. Protons as well as Na^+ or K^+ will be attracted to the grooves, and the rise in local pH may be important in titrating incoming ligands, or in promoting reactions involving acidic conditions [143].

The extent of binding of proteins, peptides, drugs, polyamines, and other oligoions to linear polyelectrolytes like DNA typically decreases quite drastically as the concentration of simple salt MX increases. Small changes in the salt concentration, even with dilute salt solutions, can easily change the apparent binding by an order of magnitude or more. The observed equilibrium constant K_{obs} shows a large, negative power dependence on the salt concentration in these cases. This behavior is relatively independent of the particular M^+ or X^- ions of the 1:1 salt. Also, the slope varies systematically with the charge on the binding species, and it also depends on the charge density of the polyion species involved.

The salt sensitivity of binding (for ligands bound principally by electrostatic interactions) can be measured by examining the slope of a plot of $\ln K_{obs}$ versus the logarithm of the salt concentration. Modern theories for polyelectrolytes make definite predictions for this slope; in particular they predict a proportionality of this slope to the net charge on the ligand, but the predictions differ in detail [3,144,145]. The binding of large and structurally complex ligands (e.g., proteins) to linear polyions (e.g., DNA) presents a considerably greater challenge to thermodynamic dissection. With these large ligand systems, the quantity $\partial \ln K_{obs}/\partial \ln [MX]$ is no longer necessarily proportional to the net charge of the complex ligand; that is, the overall charge on, say, a protein cannot be identified with any certainty as the effective charge involved in counterion displacement. A protein that is overall electrically neutral, or even one that carries a net negative charge, may bind to DNA as if it were a ligand with a net positive charge. For these complicated systems, the effective charge with respect to the polyelectrolyte effect must be limited to charged groups on the ligand at the interface with the polyion. Furthermore, the observed binding constant K_{obs} and its apparent salt dependence $\partial \ln K_{obs}/\partial \ln [MX]$ may contain appreciable contributions from coion binding and from hydration effects that can augment or oppose the effects of counterion release.

1.5.2 NUCLEIC ACID DOUBLE HELICES: CONTACTS IN THE GROOVES

In both the A and B form helices, there are the complementary, or Watson-Crick, pairing of bases: adenine with thymine (or uracil, in the case of RNA), and guanine with cytosine. The bases in these pairs are held coplanar through hydrogen bonding; this planarity permits their characteristic stacking. The two complementary strands are aligned in antiparallel fashion, with the bases toward the interior and the sugar-phosphate backbone exposed to solvent. The backbone and the base pairs wind around in a right-handed double helix. The helix forms two very noticeable grooves whose walls are sugar-phosphate residues, and whose floors are formed from the edges of the bases. In the B form helix the major groove averages 11.6 Å across, and is about 8.5 Å deep; the minor groove is about 6 Å across and 8.2 Å deep, appreciably narrower and deeper than the other groove [146]. Molecular modeling indicates that on average 171 Å^2 of surface is solvent accessible, per base pair, with 44% of this area attributable to charged groups, 2.5% to polar amide groups, 16% to other polar groups, and 38% to nonpolar groups [147]. The groove width can, however, vary with base sequence, so that for individual base pairs the distribution of polarity and the total solvent-accessible area will depend on the exact sequence. The grooves differ in terms of electrostatics; the minor groove of the B form has a more negative potential than the major groove, with AT-rich sequences having the greatest negativity.

The A form helix has grooves that are qualitatively different from those of the B form; the minor groove is shallow and broad while the major groove is narrow and deep. The minor groove is sufficiently large and flexible to accommodate protein α-helices, single oligopeptide strands, and even the two strands of oligopeptides with hairpin turns. The major groove of the A form helix is, however, generally too narrow to accommodate α-helices or two strands of a β-sheet [92,148,149].

The sugar-phosphate backbone offers many van der Waals, H-bond, and ionic contacts. However, it does not change its features very much as the underlying base sequence varies. It appears that interactions with the backbone are good mostly for nonspecific association and for stabilizing (but not necessarily selecting) sequence-specific complexes. On the other hand, in the B, Z, and A form helices, the edges of the stacked base pairs face the grooves and are exposed to solvent. These edges offer multiple points for hydrogen bonding, as well as for hydrophobic contacts (e.g., the methyl group on thymine). They provide enough of a network of contacts to enable specific sequences to be recognized, as discussed in Section 1.5.4.

1.5.3 INTERCALATIVE BINDING

Lerman [150] first proposed this mode of binding to describe the interaction of DNA with acridine derivatives, but it has since become apparent that many other planar aromatic dye and drug molecules bind to nucleic acids this way as well. Characteristic features of an intercalative complex are (1) extension of the DNA helix, (2) local unwinding of the helix, and (3) insertion of the ligand such that the plane of the intercalated ring is parallel to the base pairs [151] (see Figure 1.11). The intercalated molecule is held between the adjacent base pairs by stacking, van der Waals, and electrostatic interactions; the aromatic intercalators insert in such a way as to try to maximize their overlap with the neighboring base pairs [151,152]. Compounds with three or four fused rings fit rather well into the DNA helix, overlapping and stacking on the base pairs without protruding much into the grooves, and they bind significantly more strongly than do compounds with only a single aromatic ring or two fused rings. Generally, simple intercalators (those without attached sugars, peptide chains, etc.) have a preference for an alternating pyrimidine-purine sequence, perhaps because the stacking at such a sequence is weaker than at a purine-purine

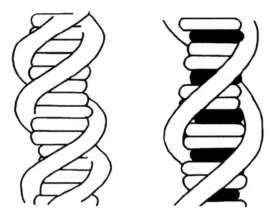

FIGURE 1.11 Intercalation into a nucleic acid double helix. Adapted with permission from *Proceedings of the National Academy of Sciences USA* Vol. 72, P.J. Bond et al., "X-ray fiber diffraction evidence for neighbor exclusion binding of a platinum metallointercalation reagent to DNA," pages 4825–4829. Copyright 1975.

sequence and so it is energetically easier to open the helix at that point for intercalation. There also appears to be a weak general preference toward intercalation of these compounds at 5′-CpG-3′ sequences versus 5′-TpA-3′ sequences.

Intercalation leaves intact the hydrogen bonding between bases in double-stranded DNA, but requires untwisting of the double helix to permit formation of a suitable space between the paired bases for insertion of the drug or dye molecule. As a result, across the intercalation site the neighboring base pairs are about twice as wide apart as normal. The degree of unwinding can, however, vary from one compound to the next; ethidium unwinds B form DNA by 26° [153,154], while acridines unwind the B form helix by 17 to 20° [155]. The unwinding extends the length of the DNA molecule and alters its buoyant density. The DNA backbone is distorted, mainly through alterations in the puckering of the ribose moiety, and there may be buckling or other distortion of the neighboring base pairs. The energetic cost of such conformational rearrangements is presumably recovered through the favorable contacts made in the final complex.

The fused ring systems of intercalators can have pendant groups that fit into the grooves and interact with DNA moieties there. These interactions can provide extra stability and sequence specificity for the intercalator. Notable examples of such compounds are the antitumor agents actinomycin D, daunorubicin, and camptothecin; the compound nogalamycin has, in fact, two such pendant groups, a nogalose moiety that fits into the minor groove and a positively charged bicyclo amino sugar that fits into the major groove [156]. Some intercalators have two intercalating moieties; for example, echinomycin has two quinoxaline rings joined by a cyclic depsipeptide. This double intercalator prefers to bind to a CpG step, with sequence recognition mediated by the depsipeptide chains making contacts in the minor groove [157].

1.5.4 SEQUENCE-SPECIFIC BINDING

A fundamental problem in molecular biology is how proteins use their amino acid residues to recognize particular DNA sequences. Diverse protein folds are used to make contact, most commonly with the backbone and major groove of the DNA [126,148]. Sequence-specific binding is at the base of many important cellular processes— gene regulation, DNA replication and recombination, and more—and the discovery of a "recognition code" that would match particular amino acids or short sequences of amino acids to particular bases or base sequences would be a great step forward toward understanding, and then manipulating, such processes. Unfortunately, no such code has been found [126,158]. Some statistical correlations have been established: Arg, Lys, Ser, and His tend to match often with guanine; Asn, Gln, His, and Pro with adenine; and Thr with thymine, the latter mainly due to the nonpolar threonine side chain matching the methyl group on thymine [126]. But researchers are still far from being able to predict beforehand which amino acid sequence would guarantee highly specific recognition of a particular DNA sequence.

Three main groups of interactions are important for protein-DNA site recognition: ionic contacts with the sugar-phosphate backbone of the DNA, van der Waals contacts with the walls and floor of the grooves of the DNA helix, and lastly, hydrogen bond contacts with groups on the walls and floors of the grooves. Since there is rather little sequence-specific variation in the features of the walls, it has been commonly assumed that any site specificity in binding must derive from a specific network of hydrogen bonds between protein and DNA, mostly along the floors of the grooves.

A survey of the crystal structures of a large number of DNA-protein complexes by Luscombe et al. [126] found that roughly two-thirds of all the contacts involved van der Waals interactions, and one-third involved hydrogen bonds. Of the hydrogen bonding interactions, about one-half of these involved an interposed water molecule; it appears these waters served less as a specific contact and more as a "filler" to avoid leaving without a partner some donor or acceptor group on the protein or DNA. Clearly, the van der Waals contacts play a major role in stabilizing the complex, though they probably do little to guarantee its sequence specificity. (Desolvation of the juxtaposed surfaces, with the entropically favorable release of the hydrating waters, may also contribute substantially [4].) As for hydrogen bond contacts, single and bidentate interactions are both common [126]. In a bidentate contact, a single residue on the protein will make two simultaneous hydrogen bonds with the DNA. For the most part, amino acid side chain atoms are used for these bidentate contacts, but main chain atoms are also found to be H-bonded in some cases.

1.5.4.1 Sequence Recognition via the Major Groove

To understand the specificity of gene-regulatory sequences, etc., consider how a ligand might discriminate or recognize sequences significantly longer than a doublet, that is, sequences that are longer than the doublet or triplet sequences recognized and bound by intercalators (even intercalators with side chains that generate part of their binding specificity). Since the walls of the helix grooves are not very informative about the DNA sequence, the obvious next place to look for specific contacts is along the floor of the grooves, over four or more base pairs. The minor groove of the B form DNA helix has very little variation in the type and spacing of possible contacts, but the major groove is much richer. For comparison, these contacts may be summarized as follows:

AT base pair floor of minor groove: N3 of adenine (acceptor), O2 on thymine (acceptor)

GC base pair floor of minor groove: N3 of guanine (acceptor), O2 on cytosine (acceptor), and N2 amino on guanine (donor)

AT base pair floor of major groove: N7 on adenine (acceptor), N6 amino on adenine (donor), O4 on thymine (acceptor), and methyl group on thymine (nonpolar)

GC base pair floor of major groove: N7 on guanine (acceptor), O6 on guanine (acceptor), and N4 of cytosine (donor)

Unfortunately, it is not possible to reliably distinguish one base pair sequence from another by making single contacts [159]. Instead, it seems there must be at least two simultaneous contacts, using H-bonds and steric/hydrophobic interactions (with the methyl group on thymine) for discrimination. There are two main alternatives here: make the simultaneous contacts along the edge of a single base or base pair, or "bridge/fork" across two adjacent bases or base pairs. Both kinds of contacts have been found on crystals of DNA-protein complexes (see Figures 1.12 and 1.13).

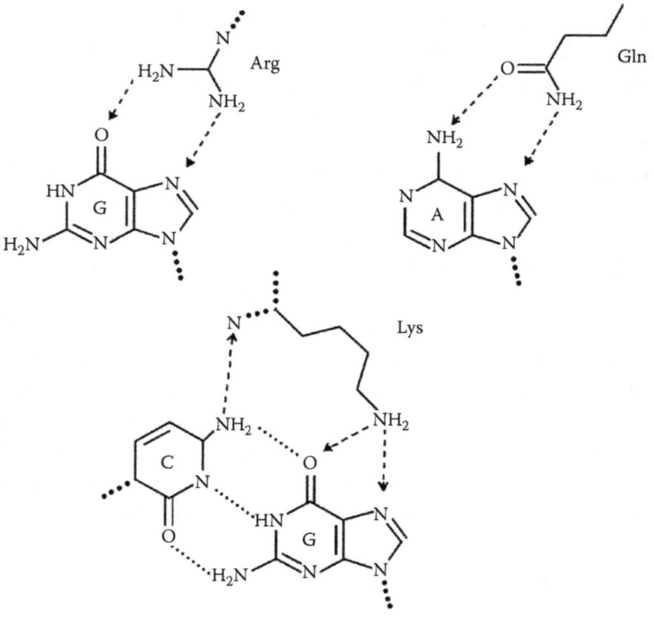

FIGURE 1.12 Bi- and tridentate interactions of amino acids with purine bases, via the major groove of the DNA helix. A. Bidentate recognition of guanine by arginine. B. Bidentate recognition of adenine by glutamine. C. Tridentate recognition of a GC base pair by lysine; note the use of the main chain residue of lysine here. From *Nucleic Acids Research* Vol. 29, N.M. Luscombe et al., "Amino acid-base interactions: A three-dimensional analysis of protein-DNA interactions at an atomic level," pages 2860–2874, by permission of Oxford University Press. Copyright 2001 Oxford University Press.

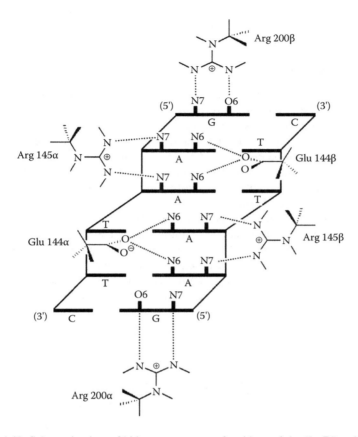

FIGURE 1.13 Schematic view of bidentate contacts of residues of the *Eco*RI endonuclease with a cognate double-stranded oligonucleotide, in the major groove. Heavy horizontal lines represent the respective DNA bases, which are connected by thinner lines representing the sugar-phosphate backbone. The hydrogen-bonding moieties on the bases that extend into the major groove are represented by the short heavy vertical lines, and their hydrogen bond contacts with the enzyme are represented by the dashed lines. The dimeric enzyme induces kinking in the backbone of the two DNA strands (indicated by the change in direction of the lines representing the sugar-phosphate backbone) and aligns hydrogen-bonding groups of adjacent adenine residues for interaction with Glu 144 and Arg 145. "Bridging" contacts with adjacent adenine N6 moieties are made by Glu 144, with similar bridging of adjacent adenine N7 moieties by Arg 145. Adapted with permission from *Science* Vol. 234, J. A. McClarin et al., "Structure of the DNA-Eco RI endonuclease recognition complex at 3 Å resolution," pages 1526–1541. Copyright 1986 AAAS.

The examples shown here emphasize the role of two simultaneous hydrogen bonds, but the methyl group on thymine should not be forgotten; it would provide a very distinctive point of contact for recognizing a thymine-adenine pair. At any rate, sequence-specific recognition at the level of single base pairs or base pair doublets can be achieved with contacts in the major groove, using as probes those functional groups that can make two (or more) simultaneous contacts. Unfortunately, there is still no well-defined "code" that would match each possible base pair doublet with a sequence of amino acids in a protein.

1.5.4.2 Site-Specific Binding in the Minor Groove

It was assumed for a long time that the minor groove of B form DNA would offer less opportunity for sequence discrimination than would the major groove, but recent work now contradicts this assumption. This reversal of viewpoint comes from studies on the binding of netropsin, distamycin, and related compounds such as Hoechst 33258 and berenil [160–162]. These compounds all fit comfortably into the minor groove of the B form DNA helix with little perturbation of DNA chain conformation, and they make hydrogen-bonding contacts with the groups on the floor of the groove (see Figure 1.14). Netropsin, distamycin, Hoechst 33258, and berenil also are positively charged, and the ionic interactions with the backbone phosphates augment their respective affinities for the DNA. These compounds prefer to bind to regions of the DNA that are rich in AT base pairs. Their binding affinity drops considerably when a GC base pair is introduced into a potential binding site; this is presumably due to the N2 amino group of guanine, which protrudes into the minor groove to interfere sterically with the compound fitting snugly into the floor of the groove.

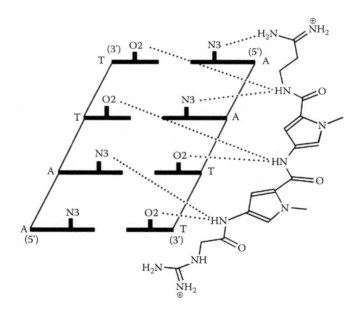

FIGURE 1.14 Schematic view of bidentate "bridging" contacts made by netropsin in the minor groove of a DNA helix, using representational conventions as in Figure 1.13 but showing hydrogen-bonding moieties in the minor groove instead. A single amide donor residue on the netropsin makes simultaneous contacts with two different residues in the groove of the helix; for example, the O2 of thymine 8 and the N3 of adenine 18 on the other strand that is part of a neighboring base pair.

Studying the crystal structure of netropsin with a double-stranded dodecamer of DNA, Kopka et al. [163] noted that replacement of one of the methylpyrrole moieties by methylimidazole would provide space for a protruding exocyclic N2 amino group of guanine, and this would give a compound capable of discriminating a GC base pair from an AT base pair at that position in the binding site. They proposed the synthesis of a series of netropsin analogs of various lengths with these substitutions at judiciously chosen points, such that the compounds would recognize and bind to specific DNA sequences containing both GC and AT base pairs. They dubbed such compounds *lexitropsins,* as derivatives of netropsin that could "read" the DNA sequence. Lown and coworkers reported soon after on just such a set of compounds [164]. It was then found that antiparallel "side-by-side" binding of two such oligoamides/peptides in the minor groove was possible [165]. The Dervan group followed this with studies on the binding of hairpin-turn oligoamides, compounds that are essentially two small lexitropsins joined by a short, flexible tether that keeps the rings of the lexitropsin moieties in proper register with each other [149]. These compounds can be designed now to recognize and bind to any particular (short) DNA sequence, using a recognition code worked out by the Dervan group [149,166]. With this code it is now possible to design hairpin polyamides that will target particular DNA sequences and affect, for example, the expression of particular genes. Obviously, such compounds could act as gene repressors; but it is also possible to use them to target adjacent sequences and promote local conformational changes, or interactions with, e.g., RNA polymerase, to promote the expression of a gene.

1.5.5 NONSPECIFIC BINDING AND LIGAND SEQUESTRATION

Studies on DNA-protein binding with various systems usually show the presence of so-called degenerate sites, with one or more bases replaced with an apparently incorrect base. These sites have (generally) lower affinity than the original or "canonical" site, yet bind the protein tightly enough to be easily distinguished from a general, nonspecific affinity of the protein for DNA. Complicating this is the statistical effect of degeneracy, in that the degenerate sequences (differing from the canonical in one, two, or more base pairs) are more likely to be encountered by chance. Even though their binding affinity may be lower than the true or canonical site, the sheer weight of numbers of these degenerate sites may overcome this, which may lead to a net uptake of ligand distributed over such sites. In fact, the degenerate sites may sequester a relatively large pool of ligand, in that the ligand is not free in solution, but is bound to these noncanonical sites. The free ligand concentration may be quite low in these cases, while the bound fraction of the ligand population may be relatively large.

Sequestration effects may also arise with certain gene regulatory proteins and the sites on DNA to which they bind. While the regulator may bind tightly and specifically to one or a few sites on a chromosome, it likely will have a generally weak affinity for DNA itself, regardless of particular sequence. Furthermore, the regulator protein may have a substantial (though somewhat reduced) affinity for sequences that are a near match to the specific regulatory site. How common are such sites in the genome, and how much would they affect binding at the specific site under physiological conditions?

Since a bacterial chromosome may have millions of base pairs, each of which could mark the start of a nonspecific binding site, it could well be that the concentration of such nonspecific sites is high enough to bind a substantial fraction of the regulator protein; that is, to reduce the overall concentration of free protein considerably. This, then, would affect binding at the specific regulatory site: if the available, free regulator protein concentration is too low, then there might not be enough to drive occupancy of the specific site on the chromosome. For example, Yang and Nash [167] have calculated that for the Integration Host Factor (IHF) protein in *E. coli*, nonspecific DNA binding reduces the free IHF concentration to 15 to 35 nM, while the total cellular IHF concentration is in the range 6 to 30 μM, three orders of magnitude larger. Similarly, von Hippel has calculated the occupancy of the *lac* operator site in *E. coli* under physiological conditions, finding that binding to secondary sites by the *lac* repressor protein has a substantial effect in reducing operator occupancy [168].

Now for the question of how common such secondary sites might be. A simple calculation for a bacterial genome containing 10 million base pairs shows that a sequence of 12 base pairs ought to arise by chance in this genome less than once [169]. That is, one can use 12 base pairs with a defined sequence to create unique sites in this bacterium. (Of course, a longer sequence would be needed for organisms with a larger genome.) However, sequences that are 10 bases in length and match 10 out of 12 bases in this unique sequence will have a frequency of $1/10^6$; that is, there will often (by chance) be 10 such sites on a chromosome with 10 million base pairs. If the affinity of the repressor protein for these sites is not far from (but of course less than) its affinity for the specific 12-base pair site, then there might be a considerable amount of binding at these noncanonical secondary sites. Such statistical considerations, along with allowances for fluctuations in protein concentrations, place limits on the range of binding constants and total concentrations of gene-regulatory proteins that can be expected in vivo [169].

REFERENCES

1. Wyman, J., Linked functions and reciprocal effects in hemoglobin: A second look, *Adv. Prot. Chem.,* 19, 223, 1964.
2. Wyman, J. and Gill, S.J., *Binding and Linkage.* Mill Valley, CA: University Science Books, 1990.
3. Record, M. T. Jr., Zhang, W., and Anderson, C.F., Analysis of effects of salts and uncharged solutes on protein and nucleic acid equilibria and processes: A practical guide to recognizing and interpreting polyelectrolyte effects, Hofmeister effects, and osmotic effects of salts, *Adv. Protein Chem.* 51, 281, 1998.
4. Spolar, R. and Record M.T. Jr., Coupling of local folding to site-specific binding of proteins to DNA, *Science* 263, 777, 1994.
5. Timasheff, S.N., Control of protein stability and reactions by weakly interacting cosolvents: The simplicity of the complicated, *Adv. Protein Chem.* 51, 355, 1998.
6. Tanford, C., *The hydrophobic Effect: Formation of Micelles and Biological Membranes*, Reprint of 2nd ed. Malabar, FL: Krieger Publishing, 1991, p. 139.
7. Fersht, A.R. et al., Hydrogen bonding and biological specificity analysed by protein engineering, *Nature* 314, 235, 1985.
8. Rose, G.D. and Wolfenden, R., Hydrogen bonding, hydrophobicity, packing, and protein folding, *Annu. Rev. Biophys. Biomol. Struct.* 22, 381, 1993.

9. Fersht, A.R., Catalysis, binding and enzyme-substrate complementarity, *Proc. R. Soc. Lond. B* 187, 397, 1974.

10. Green, N.M., Thermodynamics of the binding of biotin and some analogues by avidin, *Biochem. J.* 101, 774, 1966.

11. Courtney, E.S. et al., Thermodynamic analysis of interactions between denaturants and protein surface exposed on unfolding: Interpretation of urea and guanidinium chloride *m*-values and their correlation with changes in accessible surface area (ASA) using preferential interaction coefficients and the local-bulk domain model, *Proteins* 41, 72, 2000.

12. Grunwald, E. and Comeford, L.L., Thermodynamic mechanisms for enthalpy-entropy compensation, in *Protein-Solvent Interactions*, Gregory, R.B., Ed. New York: Marcel Dekker, 1995, p. 421.

13. Liu, L. and Guo, Q.-X., Isokinetic relationship, isoequilibrium relationship, and enthalpy-entropy compensation, *Chem. Rev.* 101, 673, 2001.

14. Gilli, P. et al., Enthalpy-entropy compensation in drug-receptor binding, *J. Phys. Chem.* 98, 1515, 1994.

15. Gilli, P. et al., Binding thermodynamics as a tool to investigate the mechanisms of drug-receptor interactions: Thermodynamics of cytoplasmic steroid/nuclear receptors in comparison with membrane receptors, *J. Med. Chem.* 48, 2026, 2005.

16. Exner, O., Entropy-enthalpy compensation and anticompensation: Solvation and ligand binding, *Chem. Commun.* 1655, 2000.

17. Krug, R.R., Hunter, W.G., and Grieger, R.A., Enthalpy-entropy compensation. 1. Some fundamental statistical problems associated with the analysis of van't Hoff and Arrhenius data, *J. Phys. Chem.* 80, 2335, 1976a.

18. Krug, R.R., Hunter, W.G., and Grieger, R.A., Enthalpy-entropy compensation. 2. Separation of the chemical from the statistical effect, *J. Phys. Chem.* 80, 2341, 1976b.

19. Krug, R.R., Hunter, W.G., and Grieger, R.A., Statistical interpretation of enthalpy-entropy compensation, *Nature* 261, 566, 1976c.

20. Sharp, K., Entropy-enthalpy compensation: Fact or artifact? *Protein Sci.* 10, 661, 2001.

21. Cornish-Bowden, A., Enthalpy-entropy compensation: A phantom phenomenon, *J. Biosci.* 27, 121, 2002.

22. Grunwald, E., *Thermodynamics of molecular species.* New York: Wiley, 1997, 120.

23. Eftink, M.R., Anusiem, A.C., and Biltonen, R.L., Enthalpy-entropy compensation and heat capacity changes for protein-ligand interactions: General thermodynamic models and data for the binding of nucleotides to ribonuclease A, *Biochemistry* 22, 3884, 1983.

24. Dunitz, J., Win some, lose some: Enthalpy-entropy compensation in weak intermolecular interactions, *Curr. Biol.* 2, 709, 1995.

25. Qian, H. and Hopfield, J.J., Entropy-enthalpy compensation: Perturbation and relaxation in thermodynamic systems, *J. Chem. Phys.* 105, 9292, 1996.

26. Cooper, A. et al., Heat does not come in different colours: Entropy-enthalpy compensation, free energy windows, quantum confinement, pressure perturbation calorimetry, solvation and the multiple causes of heat capacity effects in biomolecular interactions, *Biophys. Chem.* 93, 215, 2001.

27. Cooper, A., Heat capacity effects in protein folding and ligand binding: A re-evaluation of the role of water in biomolecular thermodynamics, *Biophys. Chem.* 115, 89, 2005.

28. Bernstein, F.C. et al., The Protein Data Bank: A computer-base archival file for macromolecular structures, *J. Mol. Biol.* 112, 535, 1977.

29. Berman, H.M. et al., The Protein Data Bank, *Nucleic Acids Res.* 28, 235, 2000.

30. Berman, H.M. et al., The Protein Data Bank, *Acta Cryst. D* 58, 899, 2002a.
31. Berman, H.M. et al., The nucleic acid database. A comprehensive relational database of three-dimensional structures of nucleic acids, *Biophys. J.* 63, 751, 1992.
32. Berman, H.M. et al., The Nucleic Acid Database, *Acta Cryst. D* 58, 889, 2002b.
33. Richards, F.M., Areas, volumes, packing, and protein structure, *Annu. Rev. Biophys. Bioeng.* 6, 151, 1977.
34. Schwarzenbach, G., Der chelateffekt, *Helv. Chim. Acta* 35, 2344, 1952.
35. Schwarzenbach, G. and Moser, P., Metallkomplexe mit polyaminen X: Mit tetrakis-(β-aminoäthyl)-äthylendiamin = "penten", *Helv. Chim. Acta* 36, 581, 1953.
36. Rebek, J. Jr., Model studies in molecular recognition, *Science* 235, 1478, 1987.
37. Cram, D.J. and Lein, G.M., Host-guest complexation. 36. Spherand and lithium and sodium ion complexation rates and equilibria, *J. Am. Chem. Soc.* 107, 3657, 1985.
38. Page, M.I. and Jencks, W.P., Entropic contributions to rate accelerations in enzymic and intramolecular reactions and the chelate effect, *Proc. Natl. Acad. Sci. USA* 68, 1678, 1971.
39. McCammon, J.A., Computer-aided molecular design, *Science* 238, 486, 1978.
40. Honig, B. and Nicholls, A., Classical electrostatics in biology and chemistry, *Science* 268, 1144, 2995.
41. Wade, R.C. et al., Electrostatic steering and ionic tethering in enzyme-ligand binding: Insights from simulations, *Proc. Natl. Acad. Sci. USA* 95, 5942, 1998.
42. Cooper, A., Thermodynamic fluctuations in protein molecules, *Proc. Natl. Acad. Sci. USA* 73, 2740, 1976.
43. Frauenfelder, H., Summary and outlook, in *Mobility and Function in Proteins and Nucleic Acids, Ciba Foundation Symposium 93*. London: Pitman, 1983, p. 329.
44. Ma, B. et al., Multiple diverse ligands binding at a single protein site: A matter of pre-existing populations, *Protein Sci.* 11, 184, 2002.
45. Hammes, G.G., Multiple conformational changes in enzyme catalysis, *Biochemistry* 41, 8221, 2002.
46. Teague, S.J., Implications of protein flexibility for drug discovery, *Nature Rev. Drug Disc.* 2, 527, 2003.
47. Rahuel, J., Priestle, J.P., and Grütter, M.G., The crystal structures of recombinant glycosylated human renin alone and in complex with a transition state analog inhibitor, *J. Struct. Biol.* 107, 227, 1991.
48. Eisenmesser, E.Z. et al., Enzyme dynamics during catalysis, *Science* 295, 1520, 2002.
49. Ghanouni, P. et al., Functionally different agonists induce distinct conformations in the G protein coupling domain of the β_2 adrenergic receptor, *J. Biol. Chem.* 276, 24433, 2001.
50. Kenakin, T., The physiological significance of constitutive receptor activity, *Trends Pharmacol. Sci.* 26, 603, 2005.
51. Costa, T. and Cotecchia, S., Historical review: Negative efficacy and the constitutive activity of G-protein-coupled receptors, *Trends Pharmacol. Sci.* 26, 618, 2005.
52. Xie, X.S., Single-molecule approach to dispersed kinetics and dynamic disorder: Probing conformational fluctuation and enzymatic dynamics, *J. Chem. Phys.* 117, 11024, 2002.
53. Ansari, A. et al., Protein states and protein quakes, *Proc. Natl. Acad. Sci. USA* 82, 500, 1985.
54. Kraut, J., Serine proteases: Structure and mechanism of catalysis, *Annu. Rev. Biochem.* 46, 331, 1977.
55. Vazquez-Laslop, N. et al., Recognition of multiple drugs by a single protein: A trivial solution of an old paradox, *Biochem. Soc. Trans.* 28, 517, 2000.
56. Sundberg, E.J. and Mariuzza, R.A., Luxury accommodations: The expanding role of structural plasticity in protein-protein interactions, *Structure* 8, R137, 2000.

57. Winkler, F.K. et al., The crystal structure of *Eco*RV endonuclease and of its complexes with cognate and non-cognate DNA fragments, *EMBO J.* 12, 1781, 1993.
58. Weber, P.C. et al., Structural origins of high-affinity biotin binding to streptavidin, *Science* 243, 85, 1989.
59. Klibanov, A.M., Enzymatic catalysis in anhydrous organic solvents, *Trends Biochem. Sci.* 14, 141, 1989.
60. Careri, G., Fasella, P., and Gratton, G., Statistical time events in enzymes: A physical assessment, *CRC Crit. Rev. Biochem.* 3, 141, 1975.
61. Careri, G., Fasella, P., and Gratton, G., Enzyme dynamics: The statistical physics approach, *Annu. Rev. Biophys. Bioeng.* 8, 69, 1979.
62. Gurd, F.R.N. and Rothgeb, T.M., Motions in proteins, *Adv. Prot. Chem.* 33, 73, 1979.
63. Alber, T. et al., The role of mobility in the substrate binding and catalytic machinery of enzymes, in *Mobility and Function in Proteins and Nucleic Acids, Ciba Foundation Symposium 93.* London: Pitman, 1983, p. 4.
64. Bennett, W.S. and Huber, R., Structural and functional aspects of domain motions in proteins, *CRC Crit. Rev. Biochem.* 15, 291, 1984.
65. Wiegand, G. and Remington, S.J., Citrate synthase: Structure, control, and mechanism, *Annu. Rev. Biophys. Biophys. Chem.* 15, 97, 1986.
66. Krause, K.L., Volz, K.W., and Lipscomb, W.N., Structure at 2.9-Å resolution of aspartate carbamoyltransferase complexed with the bisubstrate analogue *N*-(phosphonacetyl)-L-aspartate, *Proc. Natl. Acad. Sci. USA* 82, 1643, 1985.
67. Verma, C.S. and Fischer, S., Protein stability and ligand binding: New paradigms from in-silico experiments, *Biophys. Chem.* 115, 295, 2005.
68. Sturtevant, J.M., Heat capacity and entropy changes in process involving proteins, *Proc. Natl. Acad. Sci. USA* 74, 2236, 1977.
69. Irikura, K.K. et al., Transition from B to Z DNA: Contribution of internal fluctuations to the configurational entropy difference, *Science* 229, 571, 1985.
70. Erickson, H.P., Co-operativity in protein-protein association. The structure and stability of the actin filament, *J. Mol. Biol.* 206, 465, 1989.
71. García, A.E. and Soumpasis, D.M., Harmonic vibrations and thermodynamic stability of a DNA oligomer in monovalent salt solution, *Proc. Natl. Acad. Sci. USA* 86, 3160, 1989.
72. Finkelstein, A.V. and Janin, J., The price of lost freedom: Entropy of bimolecular complex formation, *Protein Eng.* 3, 1, 1989.
73. Tidor, B. and Karplus, M., The contribution of vibrational entropy to molecular association. The dimerization of insulin, *J. Mol. Biol.* 238, 405, 1994.
74. Ben-Tal, N. et al., Association entropy in adsorption processes, *Biophys. J.* 79, 1180, 2000.
75. Miller, D.W. and Agard, D.A., Enzyme specificity under dynamic control: A normal mode analysis of α-lytic protease, *J. Mol. Biol.* 286, 267, 1999.
76. Nuutero, S. et al., The amplitude of local angular motion of purines in DNA in solution, *Biopolymers* 34, 463, 1994.
77. Barkley, M.D. and Zimm, B.H., Theory of twisting and bending of chain macromolecules; analysis of the fluorescence depolarization of DNA, *J. Chem. Phys.* 70, 2991, 1979.
78. Frank-Kamenetskii, M.D., Fluctuational motility of DNA, in *Structure & Motion: Membranes, Nucleic acids & Proteins*, Clementi, E. et al., Eds. Guilderland, NY: Adenine Press, 1985, p. 417.
79. Berman, H.M. and Young, P.R., The interaction of intercalating drugs with nucleic acids, *Annu. Rev. Biophys. Bioeng.* 10, 87, 1981.

80. Quigley, G.J. et al., Non-Watson-Crick G·C and A·T base pairs in a DNA-antibiotic complex, *Science* 232, 1255, 1986.
81. Li, H.J. and Crothers, D.M., Relaxation studies of the proflavine-DNA complex: The kinetics of an intercalation reaction, *J. Mol. Biol.* 39, 461, 1969.
82. Chaires, J.B., Biophysical chemistry of the daunomycin-DNA interaction, *Biophys. Chem.* 35, 191, 1990.
83. Fox, K.R. and Waring, M.J., Evidence of different binding sites for nogalamycin in DNA revealed by association kinetics, *Biochim. Biophys. Acta* 802, 162, 1984.
84. Schultz, S.C., Shields, G.C., and Steitz, T.A., Crystal structure of a CAP-DNA complex: The DNA is bent by 90°, *Science* 253, 1001, 1991.
85. Erie, D.A. et al., DNA bending by Cro protein in specific and nonspecific complexes: Implications for protein site recognition and specificity, *Science* 266, 1562, 1994.
86. Kim, J.L., Nikolov, D.B., and Burley, S.K., Co-crystal structure of TBP recognizing the minor groove of a TATA element, *Nature* 365, 520, 1993.
87. Richmond, T.J. et al., Structure of the nucleosomal core particle at 7 Å resolution, *Nature* 311, 532, 1984.
88. Schleif, R., DNA looping, *Annu. Rev. Biochem.* 61, 199, 1992.
89. Rippe, K., von Hippel, P.H., and Langowski, J., Action at a distance: DNA-looping and initiation of transcription, *Trends Biochem. Sci.* 20, 500, 1995.
90. Pérez-Martín, J. and de Lorenzo, V., Clues and consequences of DNA bending in transcription, *Annu. Rev. Microbiol.* 51, 593, 1997.
91. Travers, A.A., Structure and function of *E. coli* promoter DNA, *CRC Crit. Rev. Biochem.* 22, 181, 1987.
92. Steitz, T.A., Structural studies of protein-nucleic acid interaction: The sources of sequence-specific binding, *Q. Rev. Biophys.* 23, 205, 1990.
93. Frederick, C.A. et al., Kinked DNA in crystalline complex with *Eco*RI endonuclease, *Nature* 309, 327, 1984.
94. McClarin, J.A. et al., Structure of the DNA-Eco RI endonuclease recognition complex at 3 Å resolution, *Science* 234, 1526, 1986.
95. Reinisch, K.M. et al., The crystal structure of HaeIII methyltransferase covalently complexed to DNA: An extrahelical cytosine and rearranged base pairing, *Cell* 82, 143, 1995.
96. O'Gara, M., Roberts, R.J., and Cheng, X., A structural basis for the preferential binding of hemimethylated DNA by *Hha*I DNA methyltransferase, *J. Mol. Biol.* 263, 597, 1996.
97. Hendlich, M. et al., Relibase: Design and development of a database for comprehensive analysis of protein-ligand interactions, *J. Mol. Biol.* 326, 607, 2003.
98. Chen, X. et al., The Binding Database: Data management and interface design, *Bioinformatics* 18, 130, 2002.
99. Roche, O., Kiyama, R., and Brooks, C.L. III, Ligand-Protein DataBase: Linking protein-ligand complex structures to binding data, *J. Med. Chem.* 44, 3592, 2001.
100. Puvanendrampillai, D. and Mitchell, J.B.O., Protein Ligand Database (PLD): Additional understanding of the nature and specificity of protein-ligand complexes, *Bioinformatics* 19, 1856, 2003.
101. Chalk, A.J. et al., PDBLIG: Classification of small molecular protein binding in the protein data bank, *J. Med. Chem.* 47, 3807, 2004.
102. Wang, R. et al., The PDBbind database: Collection of binding affinities for protein-ligand complexes with known three-dimensional structures, *J. Med. Chem.* 47, 2977, 2004.
103. Chothia, C., Structural invariants in protein folding, *Nature* 254, 304, 1975.
104. Rose, G.D. et al., Hydrophobicity of amino acid residues in globular proteins, *Science* 229, 834, 1985.

105. Miller, S. et al., Interior and surface of monomeric proteins, *J. Mol. Biol.* 196, 641, 1987.
106. Villar, H.O. and Kauvar, L.M., Amino acid preferences at protein binding sites, *FEBS Lett.* 349, 125, 1994.
107. Bartlett, G.J. et al., Analysis of catalytic residues in enzyme active sites, *J. Mol. Biol.* 324, 105, 2002.
108. Laskowski, R.A. et al., Protein clefts in molecular recognition and function, *Protein Sci.* 5, 2438, 1996.
109. Janin, J. and Chothia, C., Role of hydrophobicity in the binding of coenzymes, *Biochemistry* 17, 2943, 1978.
110. Pettit, F.K. and Bowie, J.U., Protein surface roughness and small molecular binding sites, *J. Mol. Biol.* 285, 1377, 1999.
111. Hajduk, P.J., Huth, J.R., and Fesik, S.W., Druggability indices for protein targets derived from NMR-based screening data, *J. Med. Chem.* 48, 2518, 2005.
112. Lo Conte, L., Chothia, C., and Janin, J., The atomic structure of protein-protein recognition sites, *J. Mol. Biol.* 285, 2177, 1999.
113. Jones, S. and Thornton, J.M., Principles of protein-protein interactions, *Proc. Natl. Acad. Sci. USA* 93, 13, 1996.
114. Li, X. et al., Protein-protein interactions: Hot spots and structurally conserved residues often locate in complemented pockets that pre-organized in the unbound states: Implications for docking, *J. Mol. Biol.* 344, 781, 2004.
115. Ofran, Y. and Rost, B., Analysing six types of protein-protein interfaces, *J. Mol. Biol.* 325, 377, 2003.
116. Bairoch, A. and Apweiler, R., The SWISS-PROT protein sequence database and its supplement TrEMBL in 2000, *Nucleic Acids Res.* 28, 45, 2000.
117. Delano, W.L., Unraveling hot spots in binding interfaces: Progress and challenges, *Curr. Opin. Struct. Biol.* 12, 14, 2002.
118. Cunningham, B.C. et al., Receptor and antibody epitopes in human growth hormone identified by homolog-scanning mutagenesis, *Science* 243, 1330, 2989.
119. Cunningham, B.C. and Wells, J.A., High-resolution epitope mapping of hGH-receptor interactions by alanine-scanning mutagenesis, *Science* 244, 1081, 1989.
120. Bogan, A.A. and Thorn, K.S., Anatomy of hot spots in protein interfaces, *J. Mol. Biol.* 280, 1, 1998.
121. Ma, B. et al., Protein-protein interactions: Structurally conserved residues distinguish between binding sites and exposed protein surfaces, *Proc. Natl. Acad. Sci. USA* 100, 5772, 2003.
122. Northrup, S.H. and Erickson, H.P., Kinetics of protein-protein association explained by Brownian dynamics computer simulation, *Proc. Natl. Acad. Sci. USA* 89, 3338, 1992.
123. Wells, J.A., Binding in the growth hormone complex, *Proc. Natl. Acad. Sci. USA* 93, 1, 1996.
124. Jones, S. et al., Protein-DNA interactions: A structural analysis, *J. Mol. Biol.* 287, 877, 1999.
125. Luscombe, N.M. et al., An overview of the structures of protein-DNA complexes, *Genome Biol.* 1, 1, 2000.
126. Luscombe, N.M., Laskowski, R.A., and Thornton, J.M., Amino acid-base interactions: A three-dimensional analysis of protein-DNA interactions at an atomic level, *Nucleic Acids Res.* 29, 2860, 2001.
127. Newman, M. et al., Structure of Bam HI endonuclease bound to DNA: Partial folding and unfolding on DNA binding, *Science* 269, 656, 1995.
128. Yang, W. and Steitz, T.A., Crystal structure of the site-specific recombinase γδ resolvase complexed with a 34 bp cleavage site, *Cell* 82, 193, 1995.

129. von Hippel, P.H., On the molecular bases of the specificity of interaction of transcriptional proteins with genome DNA, in *Biological Regulation and Development,* Vol. 1, Goldberger, R.F., Ed. New York: Plenum, 1979, p. 279.

130. Rich, A., Three-dimensional structure and biological function of transfer RNA, *Acct. Chem. Res.* 10, 388, 1977.

131. Schimmel, P.R., Söll, D., and Abelson, J.N., *Transfer RNA: Structure, Properties, and Recognition.* Cold Spring Harbor, NY: Cold Spring Harbor Laboratory Press, 1979.

132. Gesteland, R.F. and Atkins, J.F., *The RNA World.* Cold Spring Harbor, NY: Cold Spring Harbor Laboratory Press, 1993.

133. Wells, R.D., Unusual DNA structures, *J. Biol. Chem.* 263, 1095, 1988.

134. Tinoco, I. Jr., Nucleic acid structures, energetics, and dynamics, *J. Phys. Chem.* 100, 13311, 1996.

135. Lebrun, A. and Lavery, R., Unusual DNA conformations, *Curr. Opin. Struct. Biol.* 7, 348, 1997.

136. Bacolla, A. and Wells, R.D., Non-B DNA conformations, genomic rearrangements, and human disease, *J. Biol. Chem.* 279, 47411, 2004.

137. Harrington, R.E. and Winicov, I. New concepts in protein-DNA recognition; Sequence-directed DNA bending and flexibility, *Progr. Nuc. Acid Res.* 47, 195, 1994.

138. Robinson, B.H. and Drobny, G.P., Site-specific dynamics in DNA: Theory, *Annu. Rev. Biophys. Biomol. Struct.* 24, 523, 1995.

139. Robinson, B.H., Mailer, C., and Drobny, G., Site-specific dynamics in DNA: Experiments, *Annu. Rev. Biophys. Biomol. Struct.* 26, 629, 1997.

140. Schellman, J.A. and Harvey, S.C., Static contributions to the persistence length of DNA and dynamic contributions to DNA curvature, *Biophys. Chem.* 55, 95, 1995.

141. Cloutier, T.E. and Widom, J., Spontaneous sharp bending of double-stranded DNA, *Mol. Cell* 14, 355, 2004.

142. Cloutier, T.E. and Widom, J., DNA twisting flexibility and the formation of sharply looped protein-DNA complexes, *Proc. Natl. Acad. Sci. USA* 102, 3645, 2005.

143. Lamm, G. and Pack, G.R., Acidic domains around nucleic acids, *Proc. Natl. Acad. Sci. USA* 87, 9033, 1990.

144. Record, M.T. Jr., Anderson, C.F., and Lohman, T.M., Thermodynamic analysis of ion effects on the binding and conformational equilibria of proteins and nucleic acids: The roles of ion association or release, screening, and ion effects on water activity, *Q. Rev. Biophys.* 11, 103, 1978.

145. Manning, G.S., The molecular theory of polyelectrolyte solutions with applications to the electrostatic properties of polynucleotides, *Q. Rev. Biophys.* 11, 179, 1978.

146. Zimmerman, S.B., The three-dimensional structure of DNA, *Annu. Rev. Biochem.* 51, 395, 1982.

147. Hong, J. et al., Preferential interactions of glycine betaine and of urea with DNA: Implications for DNA hydration and for effects of these solutes on DNA stability, *Biochemistry* 43, 14744, 2004.

148. Garvie, C.W. and Wolberger, C., Recognition of specific DNA sequences, *Mol. Cell* 8, 937, 2001.

149. Dervan, P.B., Molecular recognition of DNA by small molecules, *Bioorg. Med. Chem.* 9, 2215, 2001.

150. Lerman, L.S., Structural considerations in the interaction of DNA and acridines, *J. Mol. Biol.* 3, 18, 1961.

151. Waring, M.J., DNA modification and cancer, *Annu. Rev. Biochem.* 50, 159, 1981.

152. Gale, E.F. et al., *The molecular basis of antibiotic action*, New York: Wiley, 1981, p. 258.

153. Wang, J.C., The degree of unwinding of the DNA helix by ethidium. I. Titration of twisted PM2 DNA molecules in alkaline cesium chloride density gradients, *J. Mol. Biol.* 89, 783, 1974.

154. Pulleyblank, D.E. and Morgan, A.R., The sense of naturally occurring superhelices and the unwinding angle of intercalated ethidium, *J. Mol. Biol.* 91, 1, 1975.

155. Jones, R.L. et al., The effect of ionic strength on DNA-ligand unwinding angles for acridine and quinoline derivatives, *Nucleic Acids Res.* 8, 1613, 1980.

156. Williams, L.D. et al., Structure of nogalamycin bound to a DNA hexamer, *Proc. Natl. Acad. Sci. USA* 87, 2225, 1990.

157. Waring, M.J., Echinomycin and related quinoxaline antibiotics, in *Molecular Aspects of Anticancer Drug-DNA Interactions,* Vol. 1, Neidle, S. and Waring, M.J., Eds. Boca Raton, FL: CRC Press, 1993, p. 213.

158. Pabo, C.O. and Nekludova, L., Geometric analysis and comparison of protein-DNA interfaces: Why is there no simple code for recognition? *J. Mol. Biol.* 301, 597, 2000.

159. Seeman, N.C., Rosenberg, J.M., and Rich, A., Sequence-specific recognition of double helical nucleic acids by proteins, *Proc. Natl. Acad. Sci.* USA 73, 804, 1976.

160. Nielsen, P.E., Sequence-selective DNA recognition by synthetic ligands, *Bioconjugate Chem.* 2, 1, 1991.

161. Geierstanger, B.H. and Wemmer, D.E., Complexes of the minor groove of DNA, *Annu. Rev. Biophys. Biomol. Struct.* 24, 463, 1995.

162. Neidle, S., DNA minor-groove recognition by small molecules, *Nat. Prod. Rep.* 18, 291, 2001.

163. Kopka, M.L. et al., The molecular origin of DNA-drug specificity in netropsin and distamycin, *Proc. Natl. Acad. Sci. USA* 82, 1376, 1985.

164. Lown, J.W. et al., Molecular recognition between oligopeptides and nucleic acids: Novel imidazole-containing oligopeptides related to netropsin that exhibit altered DNA sequence specificity, *Biochemistry* 25, 7408, 1986.

165. Pelton, J.G. and Wemmer, D.E., Structural characterization of a 2:1 distamycin A·d(CGAAATTGGC) complex by two-dimensional NMR, *Proc. Natl. Acad. Sci. USA* 86, 5723, 1989.

166. Kielkopf, C.L. et al., A structural basis for the recognition of AT and TA base pairs in the minor groove of B-DNA, *Science* 282, 111, 1998.

167. Yang, S.-W. and Nash, H.A., Comparison of protein binding to DNA *in vivo* and *in vitro*: Defining an effective intracellular target, *EMBO J.* 14, 6292, 1995.

168. von Hippel, P.H. et al., Non-specific DNA binding of genome regulating proteins as a biological control mechanism: I. The *lac* operon: Equilibrium aspects, *Proc. Natl. Acad. Sci. USA* 71, 4808, 1974.

169. von Hippel, P.H. and Berg, O.G., On the specificity of DNA-protein interactions, *Proc. Natl. Acad. Sci. USA* 83, 1608, 1986.

2 Binding Isotherms

2.1 SOME DEFINITIONS AND CONVENTIONS ON NOTATION

There are several different sets of symbols in current use to describe macromolecular binding systems. The set used here may differ in some details from older published work, but should still be intuitively clear and mathematically correct.

2.1.1 THE TWO PARTNERS: LIGAND AND MACROMOLECULE

Throughout the book the symbol M will stand for the *macromolecule*, which may be an enzyme, a membrane-bound protein, a DNA molecule, etc. The term *receptor* will be used interchangeably with *macromolecule*; this is not quite the same usage as in pharmacology, in which *receptor* is generally taken to mean a membrane-bound protein that is the target of drug action. The present usage is intentionally more general, so that a receptor may also be an enzyme, a transport protein, a structural protein, a nucleic acid, etc. The symbol L represents the *ligand*, the smaller partner in the binding equilibrium. The ligand may be a drug molecule, a simple inorganic ion like Na^+, a cofactor for an enzyme, a protein binding to DNA, etc. Taking M as the larger partner in the equilibrium, M may carry one or more *binding sites* for the ligand L, regions where a ligand may be sufficiently localized so as to distinguish it from other ligands that move independently of the macromolecule. The precise form and composition of these binding sites are not the main concern here (these are problems of structure, not thermodynamics or kinetics, and Chapter 1 discussed such factors in general). Of course, when M and L are of comparable size, the distinction between ligand and macromolecule breaks down and one may arbitrarily designate one or the other species as the macromolecule.

Different species of ligand will generally be indicated as L_A, L_B, L_C, and so on, or as L_1, L_2, L_3, etc. Also, the character W will indicate water as a ligand species, X^- will indicate an inorganic anion as a ligand, and M^+ will indicate an inorganic cation; the context should make clear the distinction between M for macromolecule and M^+ for small ligating cation. Occasionally *A*, *B*, etc., may appear as subscripts on the symbol M to indicate different conformations of a macromolecule, e.g., M_A and M_B will denote different conformations of the same receptor.

It is common practice in publications to denote the uncomplexed macromolecule as M or M_0 and the one-to-one complex of macromolecule and ligand as ML; complexes with 2, 3, or *s* ligands are usually denoted as ML_2, ML_3, or ML_s. This notation is clear and easily written, but one runs into trouble when trying to use it to describe complexes in which multiple species of ligand are binding simultaneously. For instance, how should one represent the complex of a macromolecule containing several molecules of ligand species A, a few more of ligand species B, and (for good measure) some tightly held water molecules, the number of which depends on how many molecules of ligand species A versus those of species B are in the complex?

These complexes can be represented using superscripts on the M symbol. Thus, for the case just above, the complex can be represented by $M^{(i_A, j_B, w_W)}$ where i_A, j_B, and w_W are the numbers of molecules of A, B, and water, respectively, in the complex. As another example, suppose there are observable salt effects on a conformational equilibrium of the macromolecule, thanks to differences in the numbers of cations and anions bound to the different conformations of the macromolecule. The symbols $M_A^{(p_{M^+}, q_{X^-})}$ and $M_B^{(s_{M^+}, t_{X^-})}$ can then be used to indicate two different macromolecular conformations with differences in complex stoichiometry.

Finally, to maintain simplicity and focus, this discussion will usually bypass cases in which there are multivalent ligands and complexes with stoichiometries like $M_m L_n$.

2.1.2 CONCENTRATIONS OF COMPONENTS

Experimental studies of macromolecular binding deal mostly with aqueous solutions of small molecules and biopolymers, and only occasionally with organic liquids and solutions. It will therefore be assumed that, unless noted explicitly, any solution discussed in this book is an aqueous solution, and the solvent species is water. For the most part, the book will use molar concentrations; exceptions will be noted as they arise. Square brackets will indicate molar concentrations; thus, [A] is the molar concentration of species A.

[L], without a subscript, will symbolize the *free* ligand molar concentration (not bound to a macromolecule), and [L]$_t$ will symbolize the *total* molar concentration of ligand, both bound and free. Similarly, the total molar concentration of the macromolecular receptor species will be denoted by [M]$_t$. The concentration of a macromolecular receptor in complex with, say, two different ligand species will be written as [$M^{(i_A, j_B)}$]. However, notice that the concentration of a receptor in conformation A is written as [M_A]. Lastly, [$M^{(0)}$] will indicate the concentration of free (unligated) macromolecular receptor, [$M^{(1)}$] the concentration of macromolecules with one ligand bound, etc. Chapter 3 will have further details in applications to multiple simultaneous binding equilibria.

Thermodynamic quantities such as chemical potential or an activity coefficient will have the pertinent species indicated by subscripts that should be intuitively clear: M for the macromolecule, W for the solvent water, L for the ligand, A for chemical species A or receptor conformation A, etc. The meaning of a particular subscript will usually be clear from its context.

2.1.3 THE AMOUNT BOUND: BINDING DENSITY AND DEGREE OF SATURATION

A key quantity in a binding study is how many ligands are bound per macromolecule. There are different ways to represent this, but the two most common are through the *degree of saturation, Y*, and the *binding density, < r >*.

The degree of saturation Y may be defined formally as the ratio of moles of ligand bound over the moles of total available binding sites. In terms of concentrations, this is

$$Y = \frac{[\text{Bound ligand}]}{[\text{Total sites}]} \tag{2.1}$$

Alternatively, Y can be thought of as the fraction of the population of available sites occupied by ligands:

$$Y = \frac{[\text{Bound sites}]}{[\text{Total sites}]} \tag{2.2}$$

Y is a pure number between zero and one, and it has no units. It is also often symbolized by the Greek letter θ.

For a simple equilibrium in which the macromolecule has a single binding site, the concentration of bound sites is simply $[\text{M}^{(1)}]$, which is also the concentration of bound or liganded macromolecules. In the above definition for Y, the quantity [Total sites] can be replaced with the concentrations of free and bound macromolecules, to obtain

$$Y = \frac{[\text{M}^{(1)}]}{[\text{M}^{(0)}] + [\text{M}^{(1)}]} \tag{2.3}$$

For binding systems in which the macromolecule has multiple binding sites, the denominator can be replaced with a sum over the concentrations of all the different forms of the macromolecule, those with 0, 1, 2, ... , n ligands bound:

$$Y = \frac{\displaystyle\sum_{i=1}^{n} i \times [\text{M}^{(i)}]}{n \times \displaystyle\sum_{i=0}^{n} [\text{M}^{(i)}]} = \frac{[\text{M}^{(1)}] + 2 \times [\text{M}^{(2)}] + 3 \times [\text{M}^{(3)}] + \cdots + n \times [\text{M}^{(n)}]}{n \times \{[\text{M}^{(0)}] + [\text{M}^{(1)}] + [\text{M}^{(2)}] + [\text{M}^{(3)}] + \cdots + [\text{M}^{(n)}]\}} \tag{2.4}$$

Notice that the sum in the numerator involves weighting of each of the species $\text{M}^{(i)}$ by its stoichiometry of ligation i. Thus, Y is an average quantity, taken over all the macromolecules present in the sample. Individual macromolecules may be completely saturated with ligand, or completely free of any bound species, but on average the fraction of available sites that contain a ligand is given by Y.

A quantity closely related to Y is the *binding density*. This is defined formally as the average number of ligands bound *per macromolecule*, or the ratio of total moles of bound ligand per mole of macromolecule (notice that this is *not* per mole of binding sites; the macromolecule may have more than one binding site but here the appropriate quantity to count is number of molecules, not sites). The binding density is a pure number with a value between zero and n, the number of binding sites per macromolecule. Common symbols for the binding density are the Greek letter ν or the Roman letter r; this book will use $< r >$, with the brackets emphasizing that this is an average quantity. The mathematical definition of the binding density in fact shows that the binding density $<r>$ is equivalent to a number average for ligands bound per macromolecule:

$$<r> = \frac{\sum_{i=1}^{n} i \times [M^{(i)}]}{\sum_{i=0}^{n} [M^{(i)}]} = \frac{[M^{(1)}] + 2 \times [M^{(2)}] + 3 \times [M^{(3)}] + \cdots + n \times [M^{(n)}]}{[M^{(0)}] + [M^{(1)}] + [M^{(2)}] + [M^{(3)}] + \cdots + [M^{(n)}]} \quad (2.5)$$

The binding density is simply related to the degree of saturation Y by

$$<r> = nY \quad (2.6)$$

Y and $<r>$ are of course identical for a macromolecule with a single binding site ($n = 1$).

The definition of binding density becomes murky when the nature of the binding site is not well defined. This is especially important in systems in which the binding is weak and there are many possible binding sites, as for example with the hydration of proteins. Confusion can also arise when dealing with linear lattices of binding sites and with large ligands that may cover up more than one lattice site. In Chapter 5, which concerns ligand binding to linear polymers like DNA, the matter of an appropriate measure of the extent of binding will have to be reconsidered for these linear systems.

Another quantity that is closely related to the binding density is the ratio of moles of ligand bound per *unit mass* of macromolecule. In pharmacology the usual symbol for this quantity is B, and it is common practice in that field to use F as the symbol for the concentration of free ligand. The mathematical definition of B is

$$B = \frac{n_{L,bound}}{m_{M,total}} \quad (2.7)$$

where $n_{L,bound}$ is the number of *moles* of bound ligand (*not* the binding stoichiometry) and $m_{M,total}$ is the total *mass* (not moles) of macromolecular receptor. The quantity B, with its mixed units, is most useful when the total mass per unit volume of macromolecule is known but not its molar concentration. This is often the case with membrane-bound receptor proteins, in which the total mass of protein per unit volume of sample is known, but the protein's molecular weight has not been determined and so its molar concentration is not known. Pharmacological studies often report the quantity B_{max}, a maximum value for B that corresponds to receptor saturation. This quantity is, however, a measure of the average number of such receptors per cell (or per microsomal vesicle, etc.), and is not the same as the number of sites per individual macromolecule.

2.1.4 Notation for Binding Constants

There is potential for confusion in the use of the letter k for both binding constants and kinetic parameters. In this book the convention will be to use an italic k, K, or K

for binding constants, and a roman k for kinetic parameters. Also, the book will occasionally use dissociation constants, usually written as K_{diss}. All these quantities will commonly be expressed on the molarity concentration scale. Finally, Boltzmann's constant will be written as k_B.

What about the units of the equilibrium constant? As the argument of a logarithm in the relation connecting free energy change to binding constant, K should be dimensionless. Still, a naive formulation of K as the concentration ratio for the reaction $M^{(0)} + L \rightleftarrows M^{(1)}$, with $K = [M^{(1)}]/\{[M^{(0)}] \cdot [L]\}$, shows that K apparently has the dimensions of reciprocal concentration. Given the common use of the molar concentration scale, this apparently means that K should have the units of M^{-1} or L/mol, and the corresponding dissociation constant should have the units of molarity. One often sees reports in the literature of binding or dissociation constants with just these units attached. It is, however, formally an error to assign units to the equilibrium constant. Just as in the relation

$$\mu_i = \mu_i^0 + RT \ln C_i \qquad (2.8)$$

where the concentration C_i actually represents a ratio of C_i to the standard state concentration C_i° (numerically equal to 1), the concentration equilibrium constant K contains all the C_i°. Formally including these standard state concentrations serves to cancel all the units of concentration so that K is left dimensionless. Nevertheless, the *numerical* value of K will depend on the units chosen for concentration, and so it is still a helpful practice to report values for K with units like L/mol attached, to ensure recognition of the appropriate concentration units.

One further point should be made here, about the difference between concentration and thermodynamic activity. In some cases it will be necessary to take notice of the possible nonideal behavior of the system and to use thermodynamic activities in place of simple concentrations. Of course, a proper formulation of the true thermodynamic equilibrium constant would in any case use activities and not concentrations. For example, for the reaction $A + B \rightleftarrows C$ the true thermodynamic equilibrium constant K_{th} is

$$K_{th} = \left(\frac{a_C}{a_A \, a_B} \right) = \left(\frac{\gamma_C}{\gamma_A \, \gamma_B} \right) \times \left(\frac{[C]}{[A][B]} \right) = \left(\frac{\gamma_C}{\gamma_A \, \gamma_B} \right) \times K_{obs} \qquad (2.9)$$

where γ_i is the activity coefficient of the i-th species, and K_{obs} denotes the experimentally observed ratio of molar concentrations. Activities are themselves dimensionless quantities, but their numerical values (and the numerical values of the corresponding activity coefficients) will depend on the units of concentration that are chosen.

Table 2.1 collects the notational conventions used in this book.

2.2 CONNECTING THE BINDING DENSITY $<r>$ WITH THE FREE LIGAND CONCENTRATION [L]

The immediate aim of a binding experiment is to determine the extent of binding (either the degree of saturation or the binding density) as a function of the concentration of free ligand, that is, the response of the system to changes in the amount of

TABLE 2.1
Conventions on Notation

Symbol	Definition
$[A]$	Concentration of A
C_B	Molar concentration of B
m_B	Molal concentration of B
γ_B	Activity coefficient of B (molarity concentration scale)
a_B	Activity of B
K_{th}	Thermodynamic equilibrium constant; independent of pH or other solute concentrations
K, K	Equilibrium constant
k	Microscopic binding constant describing ligand binding to a site on a macromolecule
K_{app}, K_{obs}	Apparent or observed equilibrium constants that may not account for all participating species or the activities of the participating species
k, k_1, k_+, etc.	Kinetic parameters (not italicized)
$\Delta G°$, $\Delta G°_{th}$	Standard Gibbs free energy change corresponding to K_{th}
ΔG_{obs}	Observed standard Gibbs free energy change corresponding to K_{obs}
ΔG_{app}	Apparent Gibbs free energy change corresponding to K_{app}
ΔS	Entropy change
ΔH	Enthalpy change

material present. The response function is known as a binding *isotherm* since the process takes place at constant temperature. The main biophysical problem is to interpret this response function in terms of numbers, classes, and affinities of binding sites. To aid in this interpretation it is most helpful to use mathematical expressions to represent the response function; these expressions will be referred to as binding isotherm equations or simply binding equations. In these expressions the free ligand concentration is usually taken as the independent variable and the dependent variable is then the system response (the binding density or the degree of saturation).

In terms of what is actually measured in the laboratory, it might seem reasonable to use the *total* amount of ligand added to the system as the independent variable. The total ligand concentration is of course the free ligand concentration plus the amount bound per unit volume. The molar amount bound depends on the concentration of macromolecule present, which means that "moles of ligand bound per liter" cannot be varied independently of [M]. Because of this, the total ligand concentration is not an independent thermodynamic quantity, so it is not a good choice of variable on which to base thermodynamic conclusions. For rigor in data interpretation, it makes better sense to use instead the equilibrium-free (or unbound) ligand concentration [L] (or better, its thermodynamic activity when dealing with concentrated solutions), even though it may not be possible to directly manipulate this free ligand concentration or activity. The free ligand concentration at equilibrium will be independent of the amount of macromolecule present, and so it is the true independent *thermodynamic* variable.

2.2.1 DESCRIBING BINDING AT THE PHENOMENOLOGICAL LEVEL

A good starting point is the simple case of reversible binding of a single species of ligand to a macromolecule. Suppose that the macromolecule has n ligand binding sites for this single species of ligand, and that the system is equilibrated so that the free ligand concentration is [L]. The macromolecules are free to exchange their bound ligands with the surrounding pool of free ligands. Some of them will have taken up just one ligand, others will have taken up two ligands, still others may have filled all n of their binding sites, and yet others may have not bound any ligands at all. Overall, there will be some sort of equilibrium distribution of the total macromolecular population over these various subpopulations. Although individual macromolecules can of course change their state of ligation, the distribution over subpopulations remains constant, so long as the free ligand concentration [L] does not change. If [L] were to increase, then by Le Chatelier's principle there should be a general uptake of ligand and an increase in the overall extent of ligand binding; also, there should be a corresponding shift in the subpopulation distribution toward those with more ligands bound per macromolecule. Conversely, if [L] were reduced, there should be a general release of bound ligands and a drop in overall extent of binding, and again a shift in the distribution but in the opposite direction. In other words, the distribution over subpopulations is a function of [L], the concentration of free ligand present. The main theoretical problem now is to relate [L] to the overall extent of ligand binding, $< r >$, which represents the average extent of binding and accounts for different contributions from the different subpopulations.

Consider the binding equilibrium by which a single macromolecule acquires its first ligand. It is not necessary to specify which particular site holds the ligand; it could be any of the n available vacant sites on the receptor. This equilibrium can be described *phenomenologically* (without reference to any particular microscopic model of the binding) with a stepwise binding constant: $K_1 = [M^{(1)}]/([M^{(0)}] [L])$. Next a second ligand attaches, and this second equilibrium can be described using a stepwise binding constant $K_2 = [M^{(2)}]/([M^{(1)}][L])$. The further addition of three, four, ... n ligands, is then described using stepwise binding constants $K_3 = [M^{(3)}]/([M^{(2)}][L])$, ... , $K_n = [M^{(n)}]/([M^{(n-1)}][L])$. These stepwise equilibrium constants (K_i for the i-th step) do not describe the binding for specific sites, but are instead averages over the many different ways one more ligand molecule may be added to a receptor macromolecule that already has $(i-1)$ ligands bound.

There is a simple expression for the fraction of macromolecules with precisely one ligand bound: $f_1 = [M^{(1)}]/[M]_t = K_1 [M^{(0)}] [L]/[M]_t$. What about expressions for the fraction of macromolecules with 2, 3, ..., n ligands bound? These are easily found to be $f_2 = [M^{(2)}]/[M]_t = K_1 \times K_2 [M^{(0)}] [L]^2/[M]_t$, $f_3 = [M^{(3)}]/[M]_t = K_1 \times K_2 \times K_3 [M^{(0)}] [L]^3/[M]_t$, and in general $f_i = [M^{(i)}]/[M]_t = K_1 \times K_2 \times K_3 \times \cdots \times K_i [M^{(0)}] [L]^i/[M]_t$. Notice the power dependence on [L] in these expressions.

The average number of ligands acquired per macromolecule (the binding density $< r >$) is a weighted sum over the concentrations of the various macromolecule sub-populations, and is given by Eq. 2.5. The numerator of this expression provides for proper weighting of the various ligation states, and the denominator is a sum over all possible states for the macromolecule, unliganded and with 1, 2, ..., n ligands bound. Substitution for the individual terms in Eq. 2.5 gives a relation for $< r >$ in terms of the free ligand concentration [L], using the stepwise equilibrium binding constants:

$$< r >= \frac{K_1[L] + 2K_1K_2[L]^2 + 3K_1K_2K_3[L]^3 + \cdots + nK_1K_2...K_n[L]^n}{1 + K_1[L] + K_1K_2[L]^2 + K_1K_2K_3[L]^3 + \cdots + K_1K_2...K_n[L]^n} \qquad (2.10)$$

This relation for the binding density was apparently first proposed by Klotz and colleagues [1,2]. The relation holds regardless of differences in affinity among the different sites or interaction among the sites. It is important to note that at this point there are no assumptions about such microscopic details, since this is a *phenomenological* description of the binding. In addition to its application to macromolecular binding systems, this level of description is widely used in descriptions of the titration of polyprotic acids, chelation of inorganic ions, etc.

There is an alternative approach to expressing $< r >$ in terms of [L], one that is closely tied to statistical thermodynamics and partition functions; it will be used freely in the rest of this book since it can be quite useful when one has a microscopic model of the binding in mind. Consider the binding of j ligands, *simultaneously*, to the receptor:

$$M^{(0)} + jL \xrightleftharpoons{K^{(j)}} M^{(j)} \qquad (2.11)$$

Note the difference in constants for this "all-at-once" binding reaction and for the stepwise addition of ligands. $K^{(j)}$ is for a reaction involving a receptor with *no* ligands bound, while $K^{(j)}$ applies to the addition of a single ligand to a receptor with $(j - 1)$ ligands already bound. It is easy to show that $K^{(j)} = K^{(1)} \times K^{(2)} \times K^{(3)} \times \cdots \times K^{(j)}$. From this, the binding density is

$$< r >= \frac{K^{(1)}[L] + 2K^{(2)}[L]^2 + 3K^{(3)}[L]^3 + \cdots + nK^{(n)}[L]^n}{1 + K^{(1)}[L] + K^{(2)}[L]^2 + K^{(3)}K_3[L]^3 + \cdots + K^{(n)}[L]^n} \qquad (2.12)$$

This expression for $< r >$ is somewhat more compact than the previous form that uses stepwise equilibrium constants, and its compactness is an advantage when treating more complicated binding systems (e.g., those with two or more types of ligand binding simultaneously). It is of course quite easy to interconvert the two methods of expression, and both forms can be found in the literature.

The polynomials in the denominators of Eqs. 2.10 and 2.12 represent an important quantity known as the *binding polynomial*, which will be symbolized by a capitalized, boldfaced **P** in a sans-serif font. The derivation and use of binding polynomials are central to the theory of macromolecular binding [3–5]. Section 2.3 will show how to write out expressions for **P** for different binding models, from which expressions relating the binding density to the free ligand concentration can be easily derived.

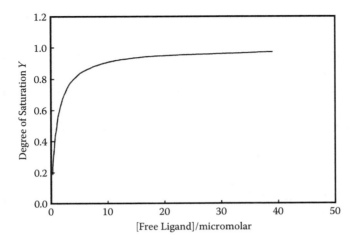

FIGURE 2.1 Direct plot of a simulated binding isotherm for a macromolecule with four independent identical sites, with site binding constant $k = 5 \times 10^6$ M^{-1}.

2.2.2 THE BINDING ISOTHERM AS A PLOT OF $< r >$ VERSUS [L]

The simplest possible graphical representation of a binding isotherm would be a plot of $< r >$ or Y on the y-axis, against the total ligand concentration on the x-axis. To compare binding models or to estimate a binding constant, however, it is proper to use the free, not total, ligand concentration for the x-axis variable, and the corresponding plot of $< r >$ or Y as a function of [L] will be referred to as a *direct plot* (Figure 2.1). For certain simple microscopic models of binding, the direct plot gives a hyperbolic curve that, as [L] increases, will asymptotically approach a maximum value for $< r >$; this maximum corresponds to complete saturation of all the available binding sites on the macromolecule. For more complicated microscopic models, other mathematical functions of greater complexity must be used to describe $< r >$ as a function of [L]. In any case, for systems with site-specific, saturable binding, the curve should eventually reach a plateau, a constant value of $< r >$, for sufficiently high values of [L].

A useful alternative way to visualize the isotherm is to plot Y versus log [L], creating a semilogarithmic or semilog plot (Figure 2.2). The result is typically a sigmoid curve whose midpoint provides an estimate of the dissociation constant for the binding system. This type of plot is commonly used in acid-base titrations and in metal-ligand complex studies, for example, and it is a popular method of representing data in competitive drug-receptor binding. The semilog plot is useful when [L] varies over two or more orders of magnitude, since it compresses the concentration scale. The extremes of very little binding and near-saturating levels of binding, which of course are found with quite different free ligand concentrations, can be shown on the same plot without crowding data points to the left side of the plot, next to the vertical axis. The semilog plot has another advantage in that it is readily

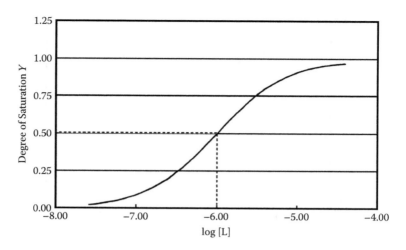

FIGURE 2.2 Determining an effective affinity from the midpoint of a titration. Simulated isotherm for a system with four equal, independent sites, with site binding constant $k = 1 \times 10^6$ M^{-1}. $[L]_{1/2}$ appears to be 1×10^6 M, indicating an effective binding constant $Q_{1/2}$ equal to 1×10^6 M^{-1}. The semilogarithmic plot compresses the concentration range and shows the approach to saturation more clearly than does a plot with a linear concentration scale (see Figure 2.1).

apparent whether or not the measurements of $< r >$ or Y approach binding saturation, to within, say, 90% of saturation. Determining (or at least estimating closely) the binding density at saturation is important for calculating the number of binding sites per macromolecule or, with linear molecules like DNA, the number of bases or base pairs involved in binding the ligand. It is common practice to use auxiliary plots like the Scatchard representation (see Section 2.4.2) for estimating these numbers. Klotz [6,7] has pointed out, however, that in these auxiliary plots it is no longer clear whether or not data have been collected near the region of binding saturation. This is a particularly ticklish point if the system has both strong and weak binding sites, where the binding to the weaker sites may be obscured or is at least not readily apparent in such plots. In such cases it is easy to be misled with respect to both the number of binding sites and the strength of the binding. This and some other points of interpreting binding curves will be considered further in Chapter 6.

2.2.3 An Effective Binding Constant: The Concentration of Free Ligand at Half Saturation

To summarize several analyses of the effects of experimental error on binding assays and related enzyme kinetic assays, it appears that the best estimates of K come from measurements when the degree of saturation Y is around 0.5 or higher [8–11]. An effective or apparent value for the binding affinity can be reached this way, without assuming any particular molecular model for the binding, as follows [12].

Experimental methods often track the equilibrium between ligand and macromolecule through a signal that is proportional to the degree of saturation. At any point along the titration there is an equilibrium between free ligand, free sites on a macromolecule, and complexed sites on a macromolecule. Because of the definition of Y, the concentration of complexed sites can be equated to $Y \times [M]_t$, and the concentration of free sites to $(1 - Y) \times [M]_t$. Then a quotient Q, analogous to an equilibrium constant, can be written as:

$$Q = \frac{Y \times [M]_t}{[L] \times (1 - Y) \times [M]_t} \tag{2.13}$$

or more simply as

$$Q = \frac{Y}{(1 - Y) \times [L]} \tag{2.14}$$

At the midpoint of the titration $Y = 0.5$, with a corresponding free ligand concentration of $[L]_{1/2}$. The midpoint value for Q, denoted as $Q_{1/2}$, is then given by

$$Q_{1/2} = \frac{1}{[L]_{1/2}} \tag{2.15}$$

This provides a convenient but crude means of characterizing the binding affinity for a system: Find the free ligand concentration corresponding to half saturation; the reciprocal of this concentration is an effective binding constant (see Figure 2.2).

Q will indeed equal the binding constant for a system with either a single binding site or with n identical binding sites; that is, Q will remain a constant for these systems, regardless of the particular value of Y or $[L]$, and it should not matter whether it is evaluated at the midpoint of the titration or at, say, 10 or 90% saturation. However, for systems in which there are multiple classes or kinds of sites, or when there are interactions among the (otherwise identical) sites, Q will vary with $[L]$ and Y; it is *not* a true equilibrium constant. In these latter cases, there will be different numerical values for Q all across the titration.

The definition and use of the quantity $Q_{1/2}$ is analogous to the use by enzymologists of "effective" Michaelis constants. For enzyme systems that do not strictly obey the Michaelis-Menten model for enzyme kinetics, an "effective" K_M can be obtained by finding the concentration of substrate that produces half the maximal velocity. A numerical value for this effective K_M can usually be determined fairly easily, and it can be useful in several different ways. For example, it can be applied to estimating substrate concentrations in which an assay for the enzyme will show appreciable activity; it may also be correlated with substrate concentrations in vivo to make predictions on the behavior of metabolic pathways. Similarly, the parameter Q is usually fairly easy to obtain, and it can be useful as a rough characterization of the binding properties of the system, for detailed planning of binding assays, for correlating with behavior in vivo, and so on.

2.2.4 THE QUESTION OF BINDING STOICHIOMETRY

From the definition of the binding density $< r >$, the upper limit to this quantity is just n, the number of specific binding sites on the macromolecule. A naive experimenter might expect that by running a titration to saturation and reading off the binding density, it would be easy to deduce the binding stoichiometry, the number of ligands held by the macromolecule at saturation. Unfortunately, matters are not so simple. To determine the stoichiometric parameter n accurately, it is necessary to make measurements under conditions in which the system is near saturation, and this may present considerable experimental difficulties. For example, achieving a degree of saturation of 99% will require about 100 times the concentration of free ligand that is needed to reach 50% saturation. If the binding is weak, the required level of free ligand may lie beyond its solubility limit. Furthermore, data in the region near binding saturation tend to be less accurate (that is, more noisy) than that collected in the region around half saturation. Also, the signal that is detected in most of the common binding methods will not give $< r >$ but instead Y, and the stoichiometric parameter n remains unknown.

So far, it has been assumed that the saturated complex of ligand with macromolecule will have the stoichiometry ML_n (alternatively written as $M^{(n)}$); that is, the ligand is univalent and the macromolecule is multivalent. This may not necessarily be the case, however, since the ligand might also be multivalent. At saturation the stoichiometry might be like $M_m L_n$ and not simply ML_m. This more complicated stoichiometry could arise, for example, with divalent antibodies (the ligands) binding to antigens on the surface of a cell, a virus, a microsomal preparation, etc. (these latter entities all playing the role of the multivalent "macromolecule" here), and forming cross-linked aggregates. This can in fact be a real problem with the quantitative interpretation of immunoassays. It may also arise with systems in which the binding of the ligand causes dimerization or aggregation of the receptor. Such is the case, for example, with the human growth hormone receptor, in which binding of the polypeptide hormone (the ligand) causes dimerization of receptor proteins (the macromolecules) that are attached to the cell membrane.

As a direct route to determining the stoichiometry one might try to isolate the intact, saturated complex, then break the isolated complex apart and quantitatively measure the moles released of each constituent. An alternative would be to measure the molecular weight of the separated complex, and compare this to that of the uncomplexed macromolecule. There are, of course, several methods for separating macromolecules, e.g., differential precipitation, gel filtration, or electrophoresis, and for determining the molecular weight of the complex, e.g., equilibrium sedimentation, light scattering, or even mass spectrometry. But for the typical small ligand and large macromolecule of biochemical systems, there may not be enough of a difference in molecular properties between "naked" and saturated macromolecule (e.g., their respective hydrodynamic size, electrophoretic mobility, or molecular weight) to settle the matter. Furthermore, it is no small feat to separate out saturated complexes from a mixture in dynamic equilibrium; it may not be possible to purify these complexes without them spontaneously dissociating on the timescale of ordinary laboratory measurements. In general, determination of the number of binding sites is more difficult than characterization of the binding affinity. In view of these difficulties, the usual course taken is to evaluate

n and K simultaneously by fitting an experimental isotherm with one or another microscopic model of the binding system. Constructing these models, and relating them to the binding isotherm equation, is the subject of the next section.

2.3 SIMPLE ISOTHERM MODELS VIA THE BINDING POLYNOMIAL

There is a very useful connection between the derivative $\partial \ln \mathbf{P} / \partial \ln[L]$ and the binding density, which stems from a fundamental relation between the binding polynomial and the overall free energy change for binding. This can be shown using a macroscopic or "thermodynamic" route to the binding polynomial, drawn from the work of John Schellman [5]. The binding polynomial can then be used to derive a binding isotherm equation.

2.3.1 BINDING FREE ENERGY CHANGES AND THE BINDING POLYNOMIAL

For simplicity suppose that the system is composed of a solution containing macromolecules (all the same type, with n binding sites on each), along with a single ligand species. At equilibrium there will be a mixture of macromolecular species with different numbers of ligands bound. A certain fraction of the macromolecules will have taken up zero ligands; others will have bound one, two, ... , up to n ligands from the solution. In other words, the solution contains a collection of $(n + 1)$ different species of macromolecule: $M^{(0)}, M^{(1)}, M^{(2)}, ..., M^{(i)}, ..., M^{(n)}$. Here $M^{(0)}$ represents the macromolecules that have no ligands bound; the superscript zero is used here for clarity. Though these different forms of the macromolecule are all present at the same time, they of course are not all at the same concentration; there will be a distribution of species fractions according to their relative free energies.

To get the total free energy change for binding, it is necessary to account for the fraction of macromolecules that has one ligand bound, the fraction that has two bound, etc. Now from elementary statistics, the average number of ligands bound per mole of macromolecule (the binding density, equivalent to the number average) is simply the number-weighted average of the various probabilities $f_1, f_2, ... , f_n$ of finding macromolecules with 1, 2, ... , n ligands bound:

$$<r> = \sum_{i=1}^{n} (i \times f_i) \qquad (2.16)$$

(The term for $i = 0$ does not contribute, of course.) The set of probabilities $\{f_i\}$ can also be interpreted as the fractions of the population of macromolecules with i ligands bound. Trying to compute all the fractions is clearly a complicated task.

To start, consider a macromolecular species with an arbitrary number i of bound ligands. Suppose this species were formed by the *simultaneous* (not stepwise) binding of i ligands onto a completely "ligand-free" macromolecule:

$$M^{(0)} + i\,L \rightleftharpoons M^{(i)}, \quad K^{(i)} = \frac{[M^{(i)}]}{[M^{(0)}][L]^i} \qquad (2.17)$$

The ligand concentration is raised to the power i, reflecting this simultaneous (and quite artificial) addition of all i ligands. Note that all this occurs in a dilute solution so that activity corrections need not be applied. Then the expression for the equilibrium constant can be rearranged to show explicitly that the concentrations $[M^{(i)}]$ are functions of the "free" macromolecule concentration $[M^{(0)}]$ and of the ligand concentration $[L]$:

$$[M^{(i)}] = K^{(i)}[M^{(0)}][L]^i \qquad (2.18)$$

To repeat, this is for binding i ligands all at once, *not* in stepwise fashion.

Clearly, this sort of expression can be written for $[M^{(1)}]$, $[M^{(2)}]$, and so on. The total macromolecular concentration is:

$$[M]_t = \sum_{i=0}^{n} [M^{(i)}] \qquad (2.19)$$

Substitution for the $[M^{(i)}]$ in the sum gives:

$$[M]_t = [M^{(0)}] \sum_{i=0}^{n} K^{(i)}[L]^i \qquad (2.20)$$

or, writing this sum out term by term:

$$[M]_t = [M^{(0)}]\left(1 + K^{(1)}[L] + K^{(2)}[L]^2 + K^{(3)}[L]^3 + \cdots + K^{(n)}[L]^n\right) \qquad (2.21)$$

The sum within the parentheses is the binding polynomial, symbolized by **P**, and

$$[M]_t = [M^{(0)}] \times \mathbf{P} \qquad (2.22)$$

where [4,5]

$$\mathbf{P} = 1 + K^{(1)}[L] + K^{(2)}[L]^2 + K^{(3)}[L]^3 + \cdots + K^{(n)}[L]^n = \sum_{i=0}^{n} K^{(i)}[L]^i \qquad (2.23)$$

It is conventional to set $K^{(0)}$ equal to one, and so for compactness the summation index i will run from zero to n.

As a first step in applying the binding polynomial approach to model building, the fraction f_i of polymer molecules with i ligands bound can be written as

$$f_i = \frac{[M^{(i)}]}{[M]_t} = \frac{K^{(i)}[L]^i}{\mathbf{P}} \qquad (2.24)$$

with cancellation of factors of $[M^{(0)}]$, top and bottom. Notice how compact the expression becomes with the use of the binding polynomial **P**.

Next, notice what happens upon forming the derivative of $\ln \mathbf{P}$ with respect to the logarithm of the ligand concentration:

$$\frac{\partial \ln \mathbf{P}}{\partial \ln[\mathrm{L}]} = \frac{1}{\mathbf{P}} \sum_{i=1}^{n} i \, K^{(i)} [\mathrm{L}]^i = \sum_{i=1}^{n} i \times f_i = <r> \tag{2.25}$$

which recovers the expression for the binding density. This formula provides a very handy way to deduce an explicit expression for the binding equation if one has in mind a specific binding model, one for which there is a binding polynomial.

To get at the free energy change for ligand binding in this model, it is convenient to define the thermodynamic state I of the system as that obtained *before* any ligand has bound to the macromolecule. Then state II of the system will be defined as that reached *after* binding equilibrium with the ligand is established, when $<r>$ moles of ligand have been taken up by the macromolecule. Schellman [5] has shown that the difference in free energy of these two states of the system is

$$\Delta G_{bind} = RT \ln \left(\frac{[\mathrm{M}^{(0)}]}{[\mathrm{M}]_t} \right) \tag{2.26}$$

(remember that $\mathbf{M}^{(0)}$ is the concentration of unliganded macromolecules, *after* establishing the equilibrium). Since $[\mathrm{M}]_t$ is just $[\mathrm{M}^{(0)}] \times \mathbf{P}$, a simple substitution gives

$$\Delta G_{bind} = -RT \ln \mathbf{P} \tag{2.27}$$

This is analogous to the familiar expression $\Delta G^0 = -RT \ln K$ for an ordinary chemical equilibrium. It should now be apparent that the binding polynomial $\mathbf{P}$ is analogous to an overall equilibrium binding constant, encompassing all the various binding states of the polymer.

As it turns out, the binding polynomial $\mathbf{P}$ is a simplified type of partition function for the system. The derivative formula in Eq. 2.25 is a natural consequence of the statistical thermodynamics of partition functions [3,5].

2.3.2 The Langmuir Isotherm Model: Binding to Equal Independent Sites

On to some particular binding models. A logical starting point is with a very simple model in which the macromolecules have a single binding site each, and the sites have identical affinities for the ligand. Writing the equilibrium as

$$\mathrm{M}^{(0)} + \mathrm{L} \rightleftharpoons \mathrm{M}^{(1)} \tag{2.28}$$

the corresponding thermodynamic equilibrium constant K_{th} is:

$$K_{th} = \frac{a_{\mathrm{M}^{(1)}}}{a_{\mathrm{M}^{(0)}} a_{\mathrm{L}}} = \left(\frac{\gamma_{\mathrm{M}^{(1)}}}{\gamma_{\mathrm{M}^{(0)}} \gamma_{\mathrm{L}}} \right) \times \left(\frac{[\mathrm{M}^{(1)}]}{[\mathrm{M}^{(0)}][\mathrm{L}]} \right) \tag{2.29}$$

where $a_{M^{(0)}}$, a_L, and $a_{M^{(1)}}$ are the thermodynamic activities of the free (unligated) macromolecule, the free ligand, and the complex, respectively.

This implicitly ignores any participation of the solvent in the equilibrium, e.g., competition of solvent with the ligand for the binding site on the macromolecule; changes in the degree of hydration of species across a binding equilibrium can indeed be an important factor in many systems. Also, for simplicity the system will be assumed to be sufficiently dilute that activity corrections are negligible, so that activity coefficients are unnecessary in the formulas.

If it were possible to simultaneously measure the concentrations of all three species, $M^{(0)}$, L, and $M^{(1)}$, then of course it would be immediately practicable to compute the binding or equilibrium constant. For most experimental situations, however, the three different species' concentrations are not measured directly. Instead, common practice is to measure (or calculate) the total amount of macromolecule and the total amount of ligand to be mixed together, and the experimental method then gives the degree of saturation corresponding to this composition of the system. This will give a single point in what is really a three-dimensional space of possible experimental results, viewing Y as a function of total macromolecule concentration $[M]_t$ and total ligand concentration $[L]_t$. More such points can be collected by varying the system's composition.

The problem then is to deduce the value of K_{obs} from the observed values of the binding saturation and the corresponding values of the free ligand concentration. Suppose that there is an excess of ligand over macromolecule. In terms of the degree of saturation Y and the free ligand concentration [L], an observed equilibrium constant can be formulated as

$$K_{obs} = \frac{Y}{(1-Y)[L]} \tag{2.30}$$

This is readily rearranged:

$$Y = \frac{K_{obs}[L]}{1+K_{obs}[L]} \tag{2.31}$$

This is a form of the famous *Langmuir binding isotherm equation*, here applied to a single binding site on a macromolecule. (Although this equation is associated with the physical chemist Irving Langmuir [13,14], it appears that the physiologist A. V. Hill derived it first [15].) This relation is easily generalized to a system of n identical binding sites that are all acting independently of one another. In terms of the binding density $< r >$, the result is:

$$< r >= \frac{nK_{obs}[L]}{1+K_{obs}[L]} \tag{2.32}$$

2.3.3 Multiple Classes of Independent Sites

What if there are two different kinds of binding sites for the same ligand? This is the sort of situation regularly faced by experimenters in describing binding to

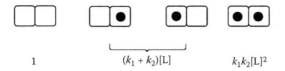

$$1 \qquad\qquad (k_1 + k_2)[\mathrm{L}] \qquad\qquad k_1 k_2 [\mathrm{L}]^2$$

FIGURE 2.3 A schematic diagram of a macromolecule with two distinct and independent binding sites (two joined hollow squares), with different microscopic binding constants k_1 and k_2, showing the different microscopic states and their statistical weights (ligand occupation indicated by solid black dots).

proteins, nucleic acids, polysaccharides, etc., and it has the potential to become a very messy exercise in algebra. Fortunately, the binding polynomial formalism can help to reduce the algebraic complications.

Consider a macromolecule with two kinds of binding sites, with one of each kind, where the sites are independent in their binding behavior but differ in their affinity for the ligand (shown schematically in Figure 2.3). For this two-site system, one general form for the binding polynomial is

$$\mathbf{P} = 1 + K_1[\mathrm{L}] + K_1 K_2 [\mathrm{L}]^2 \tag{2.33}$$

This form for $\mathbf{P}$ is implicit in common treatments of acid-base equilibria for simple amino acids, for example. It shows a quadratic dependence on [L] that reflects the presence of two sites; here the quantities K_1 and K_2 are equilibrium constants for the sequential or stepwise addition of ligands to the macromolecule:

$$\begin{aligned}
\mathrm{M}^{(0)} + \mathrm{L} &\rightleftharpoons \mathrm{M}^{(1)}, \quad K_1[\mathrm{M}^{(0)}][\mathrm{L}] = [\mathrm{M}^{(1)}] \\
\mathrm{M}^{(1)} + \mathrm{L} &\rightleftharpoons \mathrm{M}^{(2)}, \quad K_2[\mathrm{M}^{(1)}][\mathrm{L}] = [\mathrm{M}^{(2)}]
\end{aligned} \tag{2.34}$$

When $K_1 \gg K_2$ a direct plot of $< r >$ as a function of log [L] will have inflection points of maximum slope at

$$\begin{aligned}
< r > &= 0.5, \quad [\mathrm{L}] = K_1^{-1} \\
< r > &= 1.5, \quad [\mathrm{L}] = K_2^{-1}
\end{aligned} \tag{2.35}$$

In this case the uptake of ligand occurs at widely separated concentrations, and the concentrations at which the inflection points fall can then reveal the equilibrium constants K_1 and K_2. If, however, K_1 is *not* much greater than K_2, then the inflection points will merge and there will be non-negligible titration of the two sites simultaneously. See Chapter 9 of Butler [16] for further mathematical details.

Another form of the binding polynomial for this system can be written in terms of the *phenomenological* equilibrium binding constants $K^{(1)}$ and $K^{(2)}$:

$$M^{(0)} + L \rightleftharpoons M^{(1)}, \quad K^{(1)}[M^{(1)}][L] = [M^{(1)}]$$

$$M^{(0)} + 2L \rightleftharpoons M^{(2)}, \quad K^{(2)}[M^{(0)}][L]^2 = [M^{(2)}]$$

(2.36)

Under this convention the binding is described as a process of forming singly ligated macromolecules, with equilibrium constant $K^{(1)}$ for ligand addition to either one of the binding sites; then forming doubly titrated macromolecules from unbound macromolecules by the uptake of two ligands *simultaneously*, with constant $K^{(2)}$ accounting for this *simultaneous* addition of two ligands to a "naked" (unligated) macromolecule. With this convention the binding density is

$$<r> = \frac{K^{(1)}[L] + 2K^{(2)}[L]^2}{1 + K^{(1)}[L] + K^{(2)}[L]^2}$$

(2.37)

$K^{(1)}$ and $K^{(2)}$ do not directly represent the intrinsic affinity of either of the two binding sites; neither of them is equal to the microscopic binding constant of an isolated binding site of either kind. However, it is possible to relate them algebraically to the microscopic constants, and to connect the observed titration curve to those underlying (microscopic) affinities for protons.

To relate $K^{(1)}$ and $K^{(2)}$ to the underlying site binding constants k_1 and k_2, consider the free energy changes for the system as it is titrated. With the assumption that the two kinds of sites are independent of one another, the free energy changes for binding to the two kinds of sites will be additive:

$$\Delta G_{total} = \Delta G_1 + \Delta G_2$$

(2.38)

where ΔG_1 and ΔG_2 are the binding free energy changes for sites 1 and 2, respectively. Using the relation between the binding polynomial and the binding free energy change, this becomes

$$\Delta G_{total} = -RT \ln \mathbf{P}_{total}$$

(2.39)

Since the two sites are independent, there will be two independent binding polynomials, one for each site, to describe its titration. These will be called $\mathbf{P}_1$ and $\mathbf{P}_2$. Then, from $\Delta G_{total} = \Delta G_1 + \Delta G_2$, it follows that

$$\Delta G_{total} = -RT \ln \mathbf{P}_1 - RT \ln \mathbf{P}_2$$

(2.40)

and from this, the relation

$$\mathbf{P}_{total} = \mathbf{P}_1 \times \mathbf{P}_2$$

(2.41)

In other words, the overall binding polynomial for independent sites is the product of two independent binding polynomials, $\mathbf{P}_1$ and $\mathbf{P}_2$.

Suppose for this system that the first kind of site has a binding constant k_1 for ligands, and the second has a binding constant k_2. These binding constants are microscopic or site binding constants; they are what would be used to characterize these sites if one were able somehow to isolate macromolecules with just the first, or just the second, kind of site. Writing out the individual binding polynomials explicitly shows

$$\mathbf{P}_1 = (1 + k_1[L]), \quad \mathbf{P}_2 = (1 + k_2[L]) \tag{2.42}$$

Then substituting these into the expression for $\mathbf{P}_{total}$, the overall binding polynomial is

$$\mathbf{P}_{total} = 1 + (k_1 + k_2)[L] + k_1 k_2 [L]^2 \tag{2.43}$$

The first phenomenological binding constant $K^{(1)}$ is the sum $(k_1 + k_2)$, and the second phenomenological binding constant $K^{(2)}$ is the product $k_1 k_2$. In terms of a partition function for ligand binding (see the Appendix), the quantity $k_1[L]$ is the *statistical weight* given to occupation of the first kind of site, and $k_2[L]$ is the statistical weight given to the second. In terms of relative free energies G_1 and G_2 for the two kinds of occupied sites, $\ln k_1[L]$ corresponds to $-G_1/RT$ and $\ln k_2[L]$ corresponds to $-G_2/RT$; the free energy of an unoccupied site is set to zero, by convention, so that its statistical weight is simply unity.

The binding equation is now easily found by forming the double-logarithmic derivative of $\mathbf{P}$ with respect to $[L]$:

$$<r> = \frac{\partial \ln \mathbf{P}_{total}}{\partial \ln[L]} = \frac{(k_1 + k_2)[L] + 2k_1 k_2 [L]^2}{1 + (k_1 + k_2)[L] + k_1 k_2 [L]^2} \tag{2.44}$$

Notice that both the numerator and the denominator contain the ligand concentration raised to the second power. Alternatively,

$$<r> = \frac{k_1[L]}{1 + k_1[L]} + \frac{k_2[L]}{1 + k_2[L]} \tag{2.45}$$

The overall binding equation is naturally the sum of the expressions for the two independent sites.

Next, consider the more general case when there are many sites of both the first and second types on the macromolecule. Let the first class of sites have n_1 sites, each with the intrinsic site binding constant k_1, and let the second class have n_2 sites, each with the intrinsic site binding constant k_2. The total number of sites is $n = n_1 + n_2$.

Suppose i ligands were added all at once. In general, there might be a number of ligands, say, a of them, distributed over the first class of sites, and a number, b, of ligands distributed over the second class. There are, of course, many different possible combinations of a and b, under the constraints that neither a nor b, nor their sum, may exceed the total number of available sites. Then with some algebra it can be shown that the total binding polynomial for this system is

$$\mathbf{P}_{total} = \mathbf{P}_1 \times \mathbf{P}_2 = (1 + k_1 [L])^{n_1} \times (1 + k_2 [L])^{n_2} \tag{2.46}$$

The expression for the binding isotherm equation here is

$$<r> = \frac{n_1 k_1 [L]}{1 + k_1 [L]} + \frac{n_2 k_2 [L]}{1 + k_2 [L]} \tag{2.47}$$

Notice that here the binding polynomials $\mathbf{P}_1$ and $\mathbf{P}_2$ differ from those for the original system that had just one site of each type; those original binding polynomials, for a single site each, are now raised to the powers n_1 and n_2, respectively. This makes sense in terms of the total binding free energy change. If the sites are independent, then one ought to be able to add up their independent contributions to the free energy change. Because of the logarithmic relationship of the binding free energy to the binding polynomial, summing the equal free energy contributions corresponds to raising to a power the matching binding polynomial.

An easy generalization is applied to the case of m different independent classes, each class with equal and independent sites, but possibly with different numbers of sites in each class. Then

$$\mathbf{P}_{total} = \mathbf{P}_1 \times \mathbf{P}_2 \times \cdots \times \mathbf{P}_m = \left(\frac{k_1 [L]}{1 + k_1 [L]} \right)^{n_1} \times \left(\frac{k_2 [L]}{1 + k_2 [L]} \right)^{n_2} \times \cdots \times \left(\frac{k_m [L]}{1 + k_m [L]} \right)^{n_m} \tag{2.48}$$

This is only for classes of sites that are independent, however. If there is any interaction between the sites, then there is no longer a simple sum of free energy contributions to ΔG_{total}. In turn, the binding polynomial becomes more complicated. It is possible, however, to introduce interactions between sites, or to include two or more species of ligand, and still retain a relatively simple form for $\mathbf{P}_{total}$. Later, in Chapters 3 and 4, there will be examples of such binding models.

Returning to independent ligand binding with multiple classes of sites, it is not hard to show that the average number of ligands bound per macromolecule in this model is

$$<r> = \frac{n_1 k_1 [L]}{1 + k_1 [L]} + \frac{n_2 k_2 [L]}{1 + k_2 [L]} + \cdots + \frac{n_m k_m [L]}{1 + k_m [L]} \tag{2.49}$$

which is, of course, just the sum of the individual binding densities for each class. Again, this is for *independent* sites; if there is interaction between sites, this formula is not valid.

It is algebraically possible to rearrange the expression for the binding equation to the form

$$<r> = \sum_{\omega = \alpha}^{\zeta} \frac{K_\omega [L]}{1 + K_\omega [L]}, \quad \zeta = \sum_{i=1}^{m} n_i \tag{2.50}$$

The number of terms in this sum equals the total number of sites, taken over all classes of binding sites. This expression, which would describe the binding curve as the sum of simple hyperbolic curves, is a popular way to decompose curved binding isotherms in the Scatchard plot format. However, it must be used with caution.

It may happen in special cases that the quantities K_ω can be partitioned into different classes, and it might even happen that the number of classes here were the same as the actual number of classes of sites, m. Then the last expression could be recast as

$$<r>= \sum_{\omega=\alpha}^{m} \frac{n_\omega K_\omega[L]}{1 + K_\omega[L]} \tag{2.51}$$

This expression looks very much like the expression derived above for independent binding to m different classes of sites. The expression could be used to fit data for a binding isotherm, and numerical values derived from the fit might then be interpreted in terms of numbers of binding sites in particular classes, and the affinity of the respective classes for the ligand. However, Klotz and Hunston have taken pains to point out the mathematical fallacy of this simplistic interpretation when the sites are *not* independent and there are interactions between the sites [7,17].

The quantities K_ω are related to the roots of the binding polynomial (they are in fact individually equal to the reciprocals of these roots), but they do *not* correspond to the stepwise stoichiometric binding constants K_1, K_2, etc., or to the phenomenological binding constants $K^{(1)}$, $K^{(2)}$, etc. They may even take on *imaginary* values (involving the square root of -1) when the sites are not fully independent but instead are allowed to interact. Only in the particular case when all the sites are truly independent can the K_ω be equated to the microscopic equilibrium constants k_1, k_2, etc.

2.4 GRAPHICAL METHODS

Personal computers and data analysis software now are so inexpensive that computerized data evaluation is standard in any binding study claiming to present an objective analysis. Computer analysis avoids the bias of "eyeball best fit" lines, giving an objective analysis of the data. Commercial programs will not only fit a chosen model to the data and provide statistics on the goodness-of-fit, but they can take into account experimental error (estimated or measured) and can weight the data accordingly. The output from these programs can then help in determining which experimental data are statistical outliers and which are probably valid, and so help in the design of further experiments. Computer analysis relieves the analyst of the tedium of numerical comparisons of how different binding models fit the data, and so makes it faster and easier to select or discard models for further analysis. Of course, a model's good statistical fit with the data does not guarantee the correctness of the model, and other evidence (crystal structures, kinetics, etc.) should be examined in seeking to confirm the model's applicability.

Most of the popular spreadsheets for personal computers can carry out quite sophisticated statistical analyses. These spreadsheet programs can also generate graphs of the data, though often one would want a different software package for creating graphs of publication quality. There are now a number of specialized graphical software packages designed especially for scientists (not for business) that can perform both the statistical analyses and the generation of publication-quality graphics. General aspects of computer-aided analysis of experimental data are discussed later, in Chapter 6, with more detail on computer-aided methods of data analysis, a discussion of model-free analysis to derive the isotherm, some considerations of experimental error, and some aspects of nonlinear curve-fitting.

Biochemical data are frequently complicated, and it is often difficult or impractical to carry out a proper analysis of experimental error and the propagation of these errors. Without reliable estimates of the errors, it is not possible to get a truly reliable computer estimate of the goodness-of-fit of a model. However, if all that is needed is a quick estimate of a binding constant, or a quick and qualitative estimate of the cooperativity of binding, then there are some classic graphical schemes to do this. These plots are, by design, easily constructed and simple to understand, and they can be valuable guides in rapidly interpreting (with due caution) the results of binding experiments. Several of them are traditional methods of presenting the data for publication. These plots do have their weaknesses, however, and it is necessary to interpret them with care.

2.4.1 VIRTUES AND WEAKNESSES OF THE DIRECT PLOT

The direct plot, $< r >$ as a function of [L], has already been discussed in Section 2.2. This graphical form has the major virtue of clearly separating the dependent variable ($< r >$ or Y) from the independent variable (presumably the free ligand concentration [L]), and so it is perhaps more "honest" in this respect than are other graphical representations of binding data. The direct plot does, however, have some faults. First, the free ligand concentration must often be calculated by subtracting the bound ligand concentration from the total ligand concentration present, since many experimental methods do not directly measure the free ligand concentration. So [L] is not truly an independent variable but is itself usually calculated from other, more primary experimental quantities. Second, it may be difficult to decide what level of binding would correspond to saturation, since data commonly become more noisy (experimental error grows) as saturation is approached. Third, the x-axis data may cover two or more orders of magnitude in free ligand concentration, so it is hard to put all the data on the same graph without producing apparent crowding of symbols at the end corresponding to low to moderate degrees of saturation. This last difficulty can be overcome by using the logarithm of [L] for the x-axis variable, to compress ranges of concentration spanning two or more orders of magnitude while providing more separation of data points at low values of Y or $< r >$.

2.4.2 LINEARIZED PLOTS

Probably the most popular type of plot for displaying drug-receptor binding data is the Scatchard plot [18]. The plot derives from a rearrangement of the binding equation (Eq. 2.32) for a system with independent, equivalent sites:

$$\frac{< r >}{[L]} = nK_{obs} - < r > K_{obs} \qquad (2.52)$$

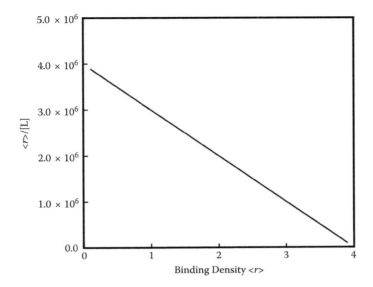

FIGURE 2.4 A Scatchard plot for a system with four identical, independent binding sites. The microscopic or site binding constant k equals 1×10^6 M^{-1}. The x-axis intercept gives the total number of sites n and the y-axis intercept gives the product nK_{obs}; the slope equals $-K_{obs}$.

This suggests a plot of $< r >/[L]$ versus $< r >$; if the model is correct then the data ought to fall on a straight line whose slope is $-K_{obs}$ and whose intercepts are nK_{obs} (on the y-axis) and n (on the x-axis). Note that the number of sites n is obtained by extrapolation of the curve to the x-axis intercept. Figure 2.4 presents an example of this type of graphical analysis.

The Scatchard plot has often been criticized for various defects. First of all, the construction of the plot does not cleanly separate the dependent variable (the binding density) from the independent variable (the ligand concentration). This results in a nonuniform propagation of errors, which makes statistical curve fitting difficult. Second, the extrapolation to the intercepts on the axes can be inaccurate if there is scatter in the data. This inaccuracy is only compounded by the usual experimental difficulty in gathering data at low or high values of the quantity ($< r >/[L]$). As the binding density rises there may be problems with getting [L] high enough to achieve near-saturation of the macromolecule; also, the experimental method may not be capable of detecting small changes in the degree of saturation under these conditions.

Finally, there is the often-encountered problem of systems with multiple sites that intrinsically cannot behave according to the equal and independent sites model. Perhaps there are two or more classes of independent binding sites, or perhaps there is interaction among the sites as they bind ligands. In such situations the data in the Scatchard plot may well follow a curve rather than a straight line, either concave-up or concave-down. In some particularly complicated systems, one can even see both types of curvature in the same plot.

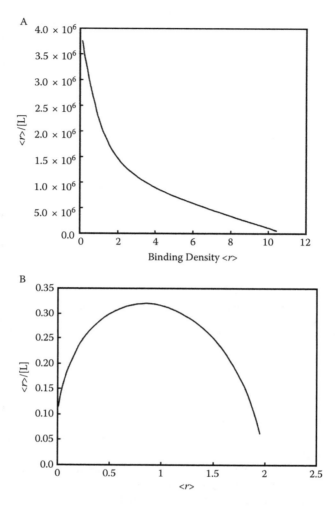

FIGURE 2.5 Curvature in Scatchard plots may be due to multiple classes of sites or to cooperativity in binding. A. Two classes of sites: a single high-affinity site ($k_1 = 3 \times 10^6\,M^{-1}$) and 10 low-affinity sites ($k_2 = 1 \times 10^5\,M^{-1}$). B. Cooperative binding in a two-site system, with $k_1 = 1 \times 10^5\,M^{-1}$, $k_2 = 1 \times 10^2\,M^{-1}$, and a cooperativity factor σ of 10^4. See Chapter 4 for an explanation of cooperativity and the factor σ.

Actually, curvature in a Scatchard plot can be quite helpful in diagnosing cases in which the simple Langmuir model of equal, independent sites does *not* apply (Figure 2.5). Plots with data trends that are concave-down (that show a "hump" in the middle) indicate positive cooperativity. Concave-up plots may indicate negative cooperativity, or they may be the result of multiple classes of independent sites (and it may be hard to distinguish between these two possibilities; more on this in Chapter 4). In either case, however, obtaining a good estimate of the number of sites is more difficult than with the standard linear plot, because the curvature of the plot makes any linear extrapolation fairly dubious.

Another linearized plot that has been commonly used to analyze binding data is the double-reciprocal plot of $1/<r>$ versus $1/[L]$. The defining equation can again be obtained by manipulating the original isotherm equation for equal and independent sites, and the result is

$$\frac{1}{<r>} = \frac{1}{n} + \frac{1}{nK_{obs}[L]} \tag{2.53}$$

If the system has just one class of equal and independent sites, then this double-reciprocal plot will be a straight line with a slope of $1/nK_{obs}$, a y-axis intercept at $1/n$, and an x-axis intercept at $-K_{obs}$. If there are multiple classes of sites or if there is cooperativity, the plot will likely be curved. As with the Scatchard plot, plot curvature with this format will make questionable any linear extrapolations to obtain numbers of sites or the binding constant.

A third linearized plot is the "half-reciprocal" plot of $[L]/<r>$ as a function of $[L]$, from the relation

$$\frac{[L]}{<r>} = \frac{1}{nK_{obs}} + \frac{[L]}{n} \tag{2.54}$$

This also is based on the model of equal and independent site binding. The plot will have an x-axis intercept at $-(1/K_{obs})$ and a y-axis intercept at $1/(nK_{obs})$. Again, curvature in the plot is indicative of the failure of the underlying assumption of equal and independent binding sites. The double-reciprocal and the half-reciprocal plots are not commonly used in the literature on ligand-binding, but their enzymological counterparts, the Lineweaver-Burk and Woolf plots, still appear in reports on enzyme kinetics.

2.4.3 Some Common Errors of Experimental Design and Interpretation

The various plots described above are designed for convenience, especially for rapid estimates of binding affinity and of the numbers of high affinity sites. Of course they will not be as accurate as a full computerized fitting of the data, and on just this basis the numbers derived from these plots should be treated rather cautiously. But the literature has many examples in which more egregious errors in data plotting and analysis have occurred. The following is a short cautionary review of some of the most common errors; Chapter 6 will have further details of data analysis and curve fitting.

2.4.3.1 Neglecting Corrections for Ligand Depletion and for Nonspecific Binding

As noted before, isotherms are best presented with the concentration of *free* ligand as the independent variable, as this is the proper thermodynamic choice. However, one often sees reports where the degree of binding is related to the *total* ligand added. Now, it is possible to rework the formulas to include the concentration of bound ligand, and so reframe them in terms of total ligand, but the underlying thermodynamics is better (and more safely) expressed in terms of the free ligand concentration. Often experiments are done under conditions in which the ligand is in vast excess over the sites, so that depletion of free ligand due to binding uptake can be neglected. Then it is a legitimate approximation to use total ligand concentration in place of the free ligand concentration. But do not forget that this is an approximation, one that becomes worse as the ratio [total ligand]/[total sites] decreases. Also, the difference between total and free ligand concentrations can be exacerbated when there is much binding of the ligand to secondary or nonspecific sites.

This brings up the question of correcting for nonspecific binding. Usually there will be some sort of weak affinity of the ligand for the macromolecule, just because the surfaces of macromolecules are so complicated and variable, and can present such a variety of possible points of attachment for a ligand. It is quite easy to find sites with weak affinity for a small organic molecule on the surface of most globular proteins (though it might be more difficult to do this with nucleic acids, due to electrostatics and less variability of the surfaces presented). Often enough, when the binding is being driven toward saturation by high ligand concentrations, there may be binding due to a second class of sites with weak affinity, and this can result in a concave-up curvature in a Scatchard plot. It is then necessary to use a more complicated model function than the simple Langmuir, one that allows for two or more classes of sites for data fitting.

Biomembranes, lipid bilayers, and micelles offer a hospitable nonpolar interior environment for small organics with nonpolar regions. Binding to receptors in such systems will often involve ligand uptake by these bilayers or micelles in addition to that bound by the receptor itself. This secondary binding is not specific, and it is typically rather weak; it is also notably difficult to saturate. The resulting experimental isotherm will then have the stronger, receptor-specific binding superimposed over this weaker but nonsaturable secondary binding; see the plot (Figure 2.6)

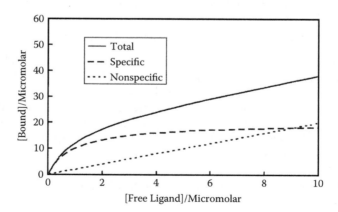

FIGURE 2.6 Bound concentration as a function of free ligand concentration. Specific binding (dashed line) compared to total binding (solid line) and to nonspecific and nonsaturable binding (dotted line), for a system with four identical, independent specific sites with binding constant $k_1 = 1 \times 10^6$ M^{-1} and nonspecific binding constant (partition coefficient) $k_2 = 2.0$, at a total receptor site concentration of 5×10^{-6} M^{-1}.

of $< r >$ as a function of [L]. Fortunately, the correction here for the nonspecific secondary binding is straightforward; the nonspecific binding can simply be viewed as equivalent to phase partitioning. Then it can be treated as a linear function of the free ligand concentration, and its contribution can be subtracted, to extract the binding that is due to specific sites.

2.4.3.2 Misinterpreting Slopes and Intercepts

It seems appropriate to focus on the Scatchard plot, since it is probably the most common data transformation used to extract binding parameters, and the one most commonly abused with respect to interpretation. Suppose that, when plotted in Scatchard format, the data appear to follow a roughly linear relationship. There will be a strong temptation to draw a straight line through the data by eye; but this must be resisted, despite the promise of a linear relation among the data. As noted above, the Scatchard plot has the weakness that it does not separate the dependent variable (the binding density) from the independent variable (the free ligand concentration). Furthermore, simple linear regression will not properly weight the data and so this procedure can lead to errors in the values of the fitted parameters (n and k). It is also not a simple matter to decide on a proper weighting of the data, since biochemical experiments are often quite complicated and it is difficult to carry out a proper propagation of error.

There is now a substantial literature on the question of errors in the determination of the binding constant [8–11,19,20]. For a system with a single binding site, the error in determining the binding constant reaches a minimum at 50% binding saturation. Also, the error climbs rapidly in the range below 20% and above 80% saturation; this trend holds qualitatively as well for systems with multiple binding sites.

Suppose that it is possible to carry out the error analysis for a particular binding system. Can one rely on the apparent goodness of the linear fit as indicating the reliability of the values for k and especially for n? In a short but vigorous discussion [6], Irving Klotz demonstrated the dangers of making this assumption by reanalyzing published data for the binding of diazepam to benzodiazepine receptors and for the binding of insulin to its receptor. The original report on benzodiazepine binding estimated the number of receptor sites at 830 per cell, by extrapolating a fitted straight line to the intercept on the x-axis. Klotz showed that the data, which seemingly covered about 85% of the entire range of binding density, actually did not reach even to half saturation of the receptors, and so the actual number of receptors per cell was at least 1400 (and likely was even higher). In the case of insulin binding to its receptor, again the original experimental data did not extend to half saturation and did not permit any accurate estimate of the number of sites. Clearly, the accuracy of the estimated stoichiometry for both systems suffered severely from inadequate data collection and overreliance on an apparently good fit.

This kind of mistake can be avoided by gathering data well past the titration midpoint (past $Y = 0.5$, on up to $Y = 0.8$ or 0.9). A simple check is to use a plot of the binding density versus the logarithm of the free ligand concentration. This plot spreads out the data more uniformly than would Scatchard or related plots, and it allows easy comparison of the degree of binding over a considerable range of free ligand concentrations. If the semilogarithmic plot shows a clear inflection point (as it should at half saturation, for equal and independent binding sites), then it is possible to make a rough estimate of the number of sites from the binding density at the midpoint. This tacitly assumes, however, that the curve continues to be a perfect sigmoid in the region where there are no observations (that is, the binding continues to obey the assumed model of equal and independent sites). The relationship to use here is simply $n = 2 <r>_{1/2}$, where $<r>_{1/2}$ is the binding density at the midpoint. Bear in mind, however, that this is an estimate based on the assumption of equal and independent sites; the relationship and thus the estimate very definitely depend on the model assumed.

Another case in which misinterpretations are easily made arises when the Scatchard plot is curved in a concave-upward fashion. The curvature indicates a decreasing affinity of the receptor for the ligand, which can stem from any one of several different sources (e.g., multiple classes of independent sites with different affinity, negative cooperative interactions among sites, a conformational change in the receptor that reduces affinity, aggregation of the receptor upon binding first ligand, with reduction in affinity for second ligand, etc.). With a curved Scatchard plot that gradually approaches the x-axis, it is very difficult to estimate the point of intersection and so to estimate the number of sites.

An example discussed by Klotz [6] is that of carbamoyl phosphate binding to aspartate transcarbamoylase, with data reported by Suter and Rosenbusch [21]. A fit of the data with the equation for a single class of independent sites gave an estimate of n equal to 4.7 ± 0.2, while a model that assumed two classes of independent binding sites, following the equation

$$<r>= \frac{n_1 k_1 [L]}{1+k_1 [L]} + \frac{n_2 k_2 [L]}{1+k_2 [L]} \tag{2.55}$$

gave $n_1 + n_2$ equal to 6.3 ± 0.5, with a statistically poorer fit but a more physically reasonable value of n for an enzyme known to possess six identical subunits. Still, if the subunits are truly identical, why should they be divided into two independent classes? A more appealing explanation, and one in line with other studies of the physical properties of the enzyme, is that the subunits cooperatively alter their conformation with a concomitant change in affinity for the ligand. The data should instead be fit with an equation that takes account of a conformational change for six subunits, a model that is a bit more complicated than those above (see Chapter 4). This points out how extra-thermodynamic information is often necessary to draw valid conclusions about numbers of sites and their classification. A single curved Scatchard plot may be fit with several different models with roughly equal (statistical) success.

In a Scatchard plot the value of the binding constant is found by extrapolation to the intercept with the y-axis. This extrapolation is rather less dangerous than the extrapolation to the x-axis intercept, since it does not require data at high binding densities that may be hard to obtain experimentally. Instead the extrapolation generally will depend on data derived at low to moderate binding densities, data that are

more easily obtained. Nevertheless, as the data in a Scatchard plot approach the
y-axis the uncertainty in the value of $< r >/[L]$ will grow; and in fact the possible
error increases dramatically as the line approaches the y-axis.

The literature has many examples of naive analyses in which the data in a
Scatchard plot are fit with two linear segments drawn through selected data at the
extremes of low and high binding density. The two-line segments are usually attrib-
uted to "low affinity" and "high affinity" sites, and their x- and y-axis intercepts are
used to deduce the numbers of binding sites and binding affinities. Now a binding
system with two classes of sites that differ appreciably in affinity ought indeed to
show a curved isotherm in a Scatchard plot, but the isotherm in the Scatchard representation cannot be simply decomposed into two tangent lines.

Klotz and Hunston [22] have shown how the limiting slopes and intercepts in
curved Scatchard (and related) plots are related to the numbers of sites and corresponding binding constants for multiple classes of independent binding sites.
Figure 2.7 shows an incorrect decomposition of the binding curve, and the necessarily more complicated analysis that is required for a multisite system. For the
case of a single ligand species binding to a receptor with m classes of sites, with the
i-th class having n_i sites, the total number of sites n_T is

$$n_T = \sum_{i=1}^{m} n_i$$

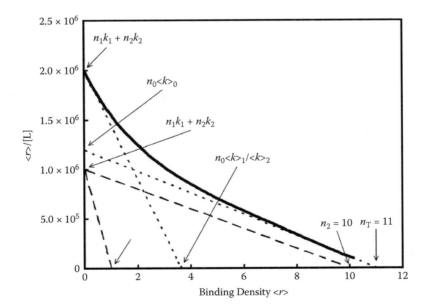

FIGURE 2.7 Incorrect decomposition of a curved Scatchard plot (thick solid line) into two
line segments (dotted lines) for a system with one high-affinity site ($k_1 = 1 \times 10^6$ M^{-1}) and
10 lower-affinity sites ($k_2 = 1 \times 10^5$ M^{-1}). The correct decomposition into two classes of sites
is indicated by the dashed lines.

and the binding density is given by

$$< r > = \sum_{i=1}^{m} \frac{n_i k_i [L]}{1 + k_i [L]} \tag{2.56}$$

From this it follows that

$$\frac{< r >}{[L]} = \sum_{i=1}^{m} \frac{n_i k_i}{1 + k_i [L]} \tag{2.57}$$

It is also convenient to define a set of average binding constants by

$$< k_i >_\gamma = \frac{\sum\limits_{i=1}^{m} n_i k_i^\gamma}{\sum\limits_{i=1}^{m} n_i k_i^{\gamma-1}} \tag{2.58}$$

where γ may be 0, 1, or 2. Now the y-axis intercept in a Scatchard plot is the limiting value of $< r >/[L]$, as both $< r >$ and $[L]$ go to zero. Klotz and Hunston find that this intercept is given by

$$\lim_{[L] \to \infty} \frac{< r >}{[L]} = n_T \times < k >_1 \tag{2.59}$$

The x-axis intercept in the plot is found by letting $[L]$ go to infinity and $< r >/[L]$ go to zero; the result is simply

$$\lim_{[L] \to 0} < r > = n_T \tag{2.60}$$

The limiting slope of the plot as it approaches the y-axis gives the (negative of the) second of the average binding constants:

$$\lim_{[L] \to 0} \frac{\partial \left(\dfrac{< r >}{[L]} \right)}{\partial < r >} = - < k >_2 \tag{2.61}$$

and as the plot approaches the x-axis, the limiting slope yields the (negative of the) zeroth average binding constant:

$$\lim_{[L] \to \infty} \frac{\partial \left(\dfrac{< r >}{[L]} \right)}{\partial < r >} = - < k >_0 \tag{2.62}$$

Note that the slopes will differ if there is more than a single class of binding sites. Klotz and Hunston analyze in detail the case for just two classes of sites; for more than two classes of sites the algebra becomes unwieldy. Klotz and Hunston give similar analyses for the other two linearized plots, e.g., $1/<r>$ as a function of $1/[L]$, and $[L]/<r>$ as a function of $[L]$.

REFERENCES

1. Klotz, I.M., The application of the law of mass action to binding by proteins. Interactions with calcium, *Arch. Biochem.* 9, 109, 1946.
2. Klotz, I.M., Walker, F.M., and Pivan, R.B., The binding of organic ions by proteins, *J. Am. Chem. Soc.* 68, 1486, 1946.
3. Hill, T.L., *Cooperativity Theory in Biochemistry.* New York: Springer-Verlag, 1985.
4. Wyman, J., Regulation in macromolecules as illustrated by haemoglobin, *Q. Rev. Biophys.* 1, 35, 1968.
5. Schellman, J.A., Macromolecular binding, *Biopolymers* 14, 999, 1975.
6. Klotz, I.M., Numbers of receptor sites from Scatchard graphs: Facts and fantasies, *Science* 217, 1247, 1982.
7. Klotz, I.M., Ligand-receptor interactions: Facts and fantasies, *Q. Rev. Biophys.* 18, 227, 1985.
8. Deranleau, D.A., Theory of the measurement of weak molecular complexes. I. General considerations, *J. Am. Chem. Soc.* 91, 4044, 1969.
9. Currie, D.J., Estimating Michaelis-Menten parameters: Bias, variance and experimental design, *Biometrics* 38, 907, 1982.
10. Bowser, M.T. and Chen, D.D.Y., Monte Carlo simulation of error propagation in the determination of binding constants from rectangular hyperbolae. 1. Ligand concentration range and binding constant, *J. Phys. Chem. A* 102, 8063, 1998.
11. Murphy, D.J., Determination of accurate KI values for tight-binding enzyme inhibitors: An in silico study of experimental error and assay design, *Anal. Biochem.* 327, 61, 2004.
12. Whitehead, E.P., Co-operativity and the methods of plotting binding and steady-state kinetic data, *Biochem. J.* 171, 501, 1978.
13. Langmuir, I., The constitution and fundamental properties of solids and liquids. Part I. Solids, *J. Am. Chem. Soc.* 38, 2221, 1916.
14. Langmuir, I., The adsorption of gases on plane surfaces of glass, mica and platinum, *J. Am. Chem. Soc.* 40, 1361, 1918.
15. Hill, A.V., The mode of action of nicotine and curari determined by the form of the concentration curve and the method of temperature coefficients, *J. Physiol.* 39, 361, 1909.
16. Butler, J.N., *Ionic Equilibrium: A Mathematical Approach.* Reading, MA: Addison-Wesley, 1964, ch. 9.
17. Klotz, I.M. and Hunston, D.L., Mathematical models for ligand-receptor binding. Real sites, ghost sites, *J. Biol. Chem.* 259, 10060, 1984.
18. Scatchard, G., The attractions of proteins for small molecules and ions, *Ann. NY Acad. Sci.* 51, 660, 1949.
19. Deranleau, D.A., Theory of the measurement of weak molecular complexes. II. Consequences of multiple equilibria, *J. Am. Chem. Soc.* 91, 4050, 1969.

20. Fuchs, H. and Gessner, R., The result of equilibrium-constant calculations strongly depends on the evaluation method used and on the type of experimental error, *Biochem. J.* 359, 411, 2001.
21. Suter, P. and Rosenbusch, J.P., Determination of ligand binding: Partial and full saturation of aspartate transcarbamylase. Applicability of a filter assay to weakly binding ligands, *J. Biol. Chem.* 251, 5986, 1976.
22. Klotz, I.M. and Hunston, D.L., Properties of graphical representations of multiple classes of binding sites, *Biochemistry* 10, 3065, 1971.

3 Binding Linkage, Binding Competition, and Multiple Ligand Species

Chapter 2 dealt with binding systems in which there was only one species of ligand present, but one or more binding sites (possibly nonidentical) per macromolecule. But the uptake or release of *multiple types* of ligands by a single macromolecule is ubiquitous throughout biochemistry. When the binding behavior of one ligand species is influenced by that of the other ligand species; there is *binding linkage* between the species.

Linkage encompasses direct competition between ligand species and cooperation in binding (through, for example, conformational changes in the macromolecule), as well as more subtle and indirect influences of one ligand species upon another (e.g., through effects on activity coefficients). It includes conformational changes in a macromolecule induced by ligand binding and "piggy-back" binding of secondary ligand species. The linkage concept can be extended to treat the effects of temperature and pressure on binding equilibria, as well as the effects of aggregational equilibria among ligands or among their macromolecular binding partners. The book by Wyman and Gill [1] gives a thorough exposition of many of these types of linkage, and more. An especially important area is the relationship between binding linkage and binding cooperativity, the subject of Chapter 4. The present chapter will be mainly concerned with *chemical linkage* involving the relatively strong binding of two or more species of ligands to a limited number of discrete sites on the macromolecule, at constant temperature and pressure.

The language of linkage thermodynamics employs a specialized vocabulary [2]. *Identical linkage* describes the direct competition of one type of ligand with another type for binding to the same site. A familiar example here would be the competition of substrate with an inhibitor for access to the active site of an enzyme. Another might be the competition of agonist with antagonist for a site on a membrane-bound drug receptor. Figure 3.1A illustrates this for the β-adrenergic receptor and some of its ligands.

Homotropic linkage describes the influence of the uptake or release of a ligand on the uptake or release of another ligand of the *same* species. An example is the binding of oxygen to the four sites in hemoglobin, where the binding of the first O_2 molecule increases the hemoglobin's affinity for subsequent molecules of O_2 (see the curve for "stripped" hemoglobin in Figure 3.1B). Homotropic linkage is closely connected to the description of cooperativity among binding sites.

Heterotropic linkage describes the influence of the uptake or release of a ligand species upon another, different ligand species. The classic example here is the Bohr

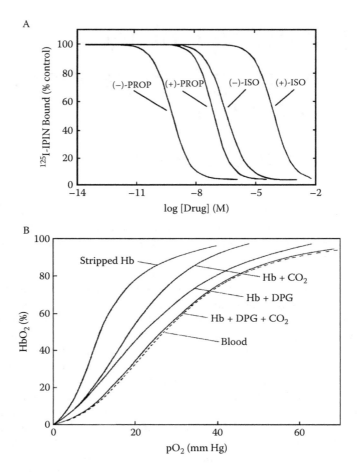

FIGURE 3.1. Examples of binding linkage. A. Identical linkage through direct binding competition. Binding of [^{125}I]-Iodopindolol to the β-adrenergic receptor is blocked and the ligand displaced by selected agonists and antagonists, as measured in a binding displacement assay. From left to right: (–)-Propranolol, (+)-Propranolol, (–)-Isoproterenol, (+)-Isoproterenol. Adapted with permission from *Journal of Pharmacology and Experimental Therapeutics* Vol. 237, M.L. Contreras, B.B. Wolfe, and P.B. Molinoff, "Thermodynamic properties of agonist interactions with the *beta* adrenergic receptor-coupled adenylate cyclase system. II. Agonist binding to soluble *beta* adrenergic receptors," pages 165–172. Copyright 1986 ASPET. B. Homotropic and heterotropic linkage through indirect, ligand-driven conformational change. Binding of a first oxygen molecule by hemoglobin cooperatively promotes further oxygen binding. Carbon dioxide or 2,3-bisphosphoglycerate (DPG) bind to hemoglobin, stabilizing it in the T ("taut" or "tense") low-affinity conformation, which reduces hemoglobin's affinity for oxygen by comparison to the "stripped" form. Adapted from *British Medical Bulletin* Vol. 32, J.V. Kilmartin, "Interaction of haemoglobin with protons, CO_2 and 2,3-diphosphoglycerate," pages 209–212. By permission of Oxford University Press; copyright Oxford University Press 1976.

effect in which solution pH (and the uptake of protons by hemoglobin) modifies the affinity of hemoglobin for O_2. Hemoglobin has other regulators of its affinity for oxygen and Figure 3.1B shows the effects of two small molecule species on the oxygenation curve.

The formal thermodynamics of binding linkage provides a unified and rigorous treatment of all these effects without invoking a microscopic model of the binding, using only the concepts of free energy or chemical potential, its dependence on concentration of components, and the mathematical properties of the free energy function and its derivatives [1–3]. However, the formal approach to linkage thermodynamics does not readily lend itself to intuitive (model-based) visualizations of linkage, precisely because it avoids microscopic models for binding. This chapter will take a less formal approach and will develop linkage concepts mainly at the microscopic level, by way of the binding polynomial. This approach, which gives a direct route to some helpful and readily applicable formulas, should also help with developing an intuitive grasp of binding linkage. A concluding section will briefly go into some of the formal thermodynamics, showing how the relations may be applied in situations in which the binding is of a diffuse and nonspecific form, such as with counterions associating with linear polyions, and cosolvent effects on the hydration of proteins.

3.1 THE BINDING POLYNOMIAL AND LINKED BINDING EQUILIBRIA

3.1.1 Positive Linkage, Negative Linkage, and No Linkage between Species

Consider a situation in which the same macromolecule can bind two different ligand species. Suppose that one wants to characterize quantitatively how these species influence one another's binding. This can be satisfied by a mathematical test, derived from the binding polynomial for the system.

Recall that the binding polynomial is related to the free energy change for uptake of free ligands by

$$\Delta G_{bind} = -RT \ln \mathbf{P} \qquad (3.1)$$

This free energy change is a function of the concentrations (actually, the thermodynamic activities) of the ligand species 1 and 2. The variation of the free energy change with those activities can be expressed in terms of its differential:

$$dG_{bind} = \left(\frac{\partial G_{bind}}{\partial \ln a_1} \right)_{a_2} d \ln a_1 + \left(\frac{\partial G_{bind}}{\partial \ln a_2} \right)_{a_1} d \ln a_2 \qquad (3.2)$$

Now, from fundamental thermodynamics, the free energy is a function of state, and as such, its differential dG_{bind} is exact. Mathematically, this is expressed as

$$\left(\frac{\partial^2 G_{bind}}{\partial \ln a_1 \, \partial \ln a_2} \right) = \left(\frac{\partial^2 G_{bind}}{\partial \ln a_2 \, \partial \ln a_1} \right) \qquad (3.3)$$

so the order of taking derivatives here is not important. The same is true of the free energy change, ΔG_{bind}, and then because ln $\mathbf{P}$ represents the free energy change upon binding, the differential d ln $\mathbf{P}$ is likewise exact. This means that

$$\left(\frac{\partial^2 \ln \mathbf{P}}{\partial \ln a_1 \, \partial \ln a_2}\right) = \left(\frac{\partial^2 \ln \mathbf{P}}{\partial \ln a_2 \, \partial \ln a_1}\right) \tag{3.4}$$

Using the derivative relation between ln $\mathbf{P}$ and binding density, it is then possible to write the differential of ln $\mathbf{P}$ as

$$d \ln \mathbf{P} = \left(\frac{\partial \ln \mathbf{P}}{\partial \ln a_1}\right)_{a_2} d \ln a_1 + \left(\frac{\partial \ln \mathbf{P}}{\partial \ln a_1}\right)_{a_1} d \ln a_2$$

$$= <r_1> d \ln a_1 + <r_2> d \ln a_2 \tag{3.5}$$

By taking cross derivatives of the binding densities $<r_1>$ and $<r_2>$, and using Eq. 3.4 one finds

$$\left(\frac{\partial <r_1>}{\partial \ln a_2}\right)_{a_1} = \left(\frac{\partial <r_2>}{\partial \ln a_1}\right)_{a_2} \tag{3.6}$$

When evaluated from binding data, the sign of one or the other of the partial derivatives here will indicate what kind of linkage it is, positive or negative. If the derivative is positive, so that the binding of one species promotes the binding of the other, then the linkage between those two species is positive. On the other hand, if the binding of one species obstructs or reduces the binding of the other, the derivative and the linkage is negative. Negative linkage of course arises when the ligands compete for binding to the same site on the receptor. However, positive or negative linkage may have more subtle sources. For example, a first ligand species might induce a conformational change in the receptor and so alter the receptor's affinity for the second species, or it might influence the solubility of the second species, changing its effective concentration (its activity) and thus the level of its binding to a receptor.

The magnitude of the derivative indicates the strength of the linkage. In particular, if the second cross derivatives of ln $\mathbf{P}$ are zero, then the binding of the two ligands is not linked at all. They bind independently of one another, having no effect on one another's binding behavior. Recall the binding polynomial for independent multiple binding by two ligand species:

$$\mathbf{P} = (1 + k_1 a_1)^n (1 + k_2 a_2)^m \tag{3.7}$$

(this allows for different numbers of sites for species 1 versus species 2). From this it is easily verified that indeed

$$\left(\frac{\partial <r_1>}{\partial \ln a_2}\right)_{a_1} = \left(\frac{\partial <r_2>}{\partial \ln a_1}\right)_{a_2} = 0 \tag{3.8}$$

This can obviously be extended to accommodate linkage among ligand species 3, 4, etc.

3.1.2 BINDING COMPETITION

Probably the simplest kind of binding linkage is the direct competition of two different species of ligand for the same site on a macromolecule. This type of linkage is seen when substrate and inhibitor compete for the active site of an enzyme, or agonist and antagonist compete for the same binding site on the receptor. A major practical application of linkage concepts arises in the pharmaceutical industry, which employs competitive binding assays quite widely to screen collections of compounds for binding activity with a particular receptor or enzyme. In many cases, the compounds, whose binding affinities need to be determined, may lack the necessary physical properties for their binding to be followed conveniently by direct assay. How then to characterize the binding of a compound for which there is not an adequate signal change? The displacement or competition assay is a standard approach in such situations.

In developing a good displacement assay, a necessary preliminary is the careful characterization of the binding of one particular species of ligand. Ideally, its binding produces a well-defined and easily detected signal that is proportional to the amount of binding; some other qualifications are listed below. This species will serve as a "probe" whose displacement from the binding site can be detected. The general procedure is then to prepare a solution of the target macromolecule with a known concentration of the probe species under conditions in which there is appreciable binding of the probe, but not to the point of binding saturation. The binding of the probe (or its release from the target) produces a suitable instrumental signal change by which to follow the binding.

For simplicity of interpretation it is best if the signal originates from the probe and not the macromolecule; perhaps the probe's fluorescence or UV absorbance could be followed, or the degree of retention of radiolabeled probe in a filtration assay. The experimenter can then simply add to the solution increasing amounts of the "unknown" whose affinity for the receptor is not yet known. If the unknown binds to the same site on the macromolecule as the probe, then the probe will eventually be displaced by the competing unknown. By following this displacement and characterizing it quantitatively, it is possible to arrive at an estimate of the affinity of the unknown for the receptor.

Ideally, the probe species and its competitors should bind to the same site(s), so that they are in direct competition with each other for those sites; also, they should not have appreciable affinity to other binding sites. This will help to keep the interpretation of the competition simple. It is also desirable, for experimental convenience, that the probe and the unknown(s) should roughly match each other in affinity, so that monitoring the probe's displacement will be a sensitive measure of the binding of the unknown. If the probe binds much more strongly than the competitor, then high levels of competitor will be needed to displace the probe; it may not be possible to achieve the necessary concentrations because of limits on solubility, aggregation, binding to secondary sites, etc. If the probe is a much weaker binder than the unknown, then the problem inverts itself; it may not be possible to displace the unknown with practicable probe concentrations. But if the affinities match, then small increases in the concentration of either species ought to have a measurable effect on the level of binding of the other species.

It is possible to use a competition assay to construct the entire binding isotherm of the competitor, and so to arrive at both n and K for this species. In many instances, however, a determination of a complete binding isotherm for a competitor is not the main goal. For example, in the pharmaceutical industry, thousands or millions of compounds may be screened against a single receptor, and all that may be sought is a quick estimate of the binding affinity of the competing compound, to determine whether or not K is high enough to make it worth further testing as a drug candidate or lead compound. In these screening assays, it is usually enough to collect data around the midpoint of the displacement titration curve, which is of course where K is most accurately determined, and then to apply some approximate relations to compare affinities.

3.1.3 A Single Class of Binding Sites and Two Competing Ligand Species

The statistical thermodynamics of competitive binding of two ligand species to the same site on a receptor macromolecule is easily developed using the binding polynomial approach. For a system with one binding site on the macromolecule, the binding polynomial is

$$\mathbf{P} = 1 + k_1[L_1] + k_2[L_2] \tag{3.9}$$

and the binding densities of the two ligand species are given by

$$<r_1> = \frac{k_1[L_1]}{1 + k_1[L_1] + k_2[L_2]}, \qquad <r_2> = \frac{k_2[L_2]}{1 + k_1[L_1] + k_2[L_2]} \tag{3.10}$$

Notice that as the concentration of competing species 2 increases, the degree of binding of species 1 declines accordingly, and vice versa; this is, of course, the basis for the ligand displacement assay.

3.1.3.1 Constant Concentration of One Species

Upon holding the (free) concentration of species 1 constant and varying that of species 2, the isotherm equation takes the form

$$<r_2> = \frac{k_2[L_2]}{A + k_2[L_2]}, \quad \text{constant } [L_1] \tag{3.11}$$

where A is a "constant" that actually depends on $[L_1]$. A plot of $<r_2>$ as a function of the logarithm of $[L_2]$ would be sigmoid, though perhaps not perfectly symmetrical, due to the possible inconstancy of A.

To try to separate out the effects of the quantity A, this equation could be rearranged to a form that suggests a half-reciprocal plot of $[L_2]/< r_2 >$ as a function of $1/[L_2]$:

$$\frac{[L_2]}{<r_2>} = [L_2] + \frac{A}{k_2} \tag{3.12}$$

Apparently the slope of the plot should be proportional, through A, to $1/k_2$.

An alternative would be to use a Hill plot of $\ln [Y/(1 - Y)]$ as a function of $\ln [L_2]$, for which the relation would be

$$\ln\left(\frac{Y_2}{1-Y_2}\right) = \ln [L_2] + \ln k_2 - \ln A \tag{3.13}$$

However, A is not a true constant, and neither of the relations is truly linear. Depending on the amount of L_1 present, the quantity A may or may not show some variation throughout the titration, possibly leading to curvature in the plot and confusion in its interpretation. Furthermore, a titration may be done with a certain initial amount of L_1 present (in order to deduce k_2 from that titration), then the experiment is repeated but with a different initial amount of L_1. In this case, it is quite possible to find rather different values for k_2. The difficulty with plot curvature (in a given titration) would be resolved if $[L_1]$ were fixed throughout the titration; the difficulty with different apparent k_2 values would be resolved if the same free concentration of L_1 were used in all titrations.

It is technically rather difficult to hold the *free* concentration of species 1 *fixed* throughout a titration, however convenient it might be for any interpretation of these formulas. It might be possible to hold the free concentration $[L_1]$ constant if L_1 were equilibrating between the solution and a second phase, say, the pure solid or liquid L_1. But this is generally not practicable for an arbitrary choice of ligands. It *is* possible to hold constant the *total* concentration of species 1; and if only a small fraction of it is bound (as with very low receptor concentrations), then it is reasonable to approximate the free concentration $[L_1]$ by that total concentration $[L_1]_t$. The slope of the half-reciprocal plot would then be unity and the intercept would be $(1 + k_1 [L_1]_t)/k_2$. The intercept would thus give a measure of the binding affinity of species 2 *relative* to that of species 1, and if k_1 were known beforehand then k_2 could be determined.

3.1.3.2 Multiple Identical Sites

Now suppose that there are n identical sites for binding per macromolecule, and the two species compete with each other at *all* these sites. This is only slightly more complicated than the simple single-site case. The binding polynomial becomes

$$\mathbf{P} = (1 + k_1[L_1] + k_2[L_2])^n \tag{3.14}$$

and the binding densities are

$$<r_1> = \frac{nk_1[L_1]}{1+k_1[L_1]+k_2[L_2]}, \quad \text{and} \quad <r_2> = \frac{nk_2[L_2]}{1+k_1[L_1]+k_2[L_2]} \tag{3.15}$$

For comparison to the single-site case, inverting the equation for $<r_2>$ gives

$$\frac{n}{<r_2>} = 1 + \frac{1+k_1[L_1]}{k_2[L_2]}$$

$$= 1 + \frac{A}{k_2[L_2]} \tag{3.16}$$

where again A is a pseudo-constant quantity that actually depends on the free concentration of L_1.

It bears repeating that the linearity of the plots derived from the relations will depend on the apparent constancy of $[L_1]$, as approximated by $[L_1]_t$. If there were any appreciable depletion of $[L_1]$, as by specific site binding or perhaps by non-specific binding to membranes, glassware, etc., then the plots would likely show curvature. It is still possible to fit the isotherm data with the depletion of species 1 explicitly taken into account, using appropriate software. Goldstein and Barrett [4] have developed guidelines for recognizing situations in which ligand depletion could lead to appreciable error and they show how, in some cases, a correction may be applied without resorting to a full nonlinear, multiparameter statistical fitting routine.

3.1.4 IC_{50} VALUES AND COMPETITION ASSAYS

When screening a number of potential ligands for a receptor in order to rank the compounds by binding affinity, a popular approach is to determine their IC_{50} values. The IC_{50} is the concentration of competitor needed to produce half inhibition of the activity of some standard ligand that is acting at a specified receptor. IC_{50} values are widely used in interpreting enzyme inhibition assays and pharmacological assays that measure receptor activation and antagonism. For pharmacological assays that are only indirect measures of the binding of ligands, these concentration values are better denoted as EC_{50} values, the effective concentration needed to produce half activation of the receptor. As regards simple binding assays for membrane-bound receptors, IC_{50} values are often determined with filtration assays in which radio-labeled probe species (whose affinities are known) compete against an unlabeled competitor whose affinity is unknown.

To determine the IC_{50} of a competitor, the total concentration of the labeled probe should be fixed, using a concentration $[L_1]_t$ chosen to produce a degree of saturation $Y_{1,max}$ that will yield a suitably large binding signal. One then titrates with competitor and follows the decrease in binding by the probe. When half of the probe is displaced by competitor (that is, upon reaching $Y_1 = 0.5 \times Y_{1,max}$ at

$[L_2] = IC_{50}$), then $k_2 \times IC_{50} = 1 + k_1 [L_1]$, so that the *dissociation* constant for the unlabeled species is simply

$$k_{2, diss} = \frac{IC_{50}}{1 + \dfrac{[L_1]}{k_{1, diss}}} \qquad (3.17)$$

This shows the proportionality between the dissociation constant for the competitor and its IC_{50}. Notice, however, that the proportionality will change upon altering the free concentration of the probe; this shows that the IC_{50} is not itself a true dissociation constant but instead depends on the details of the assay. For comparisons across different assay conditions it is preferable to calculate and report the dissociation constant instead. However, when assaying a series of competitors, the titration midpoint concentrations (the IC_{50} values), can be used to rank order the competitors by their affinity, so long as $[L_1]$ is approximately constant. Usually this means designing the assay conditions so that the free probe concentration $[L_1]$ is approximately equal to its total concentration $[L_1]_t$. This can be done by using receptor concentrations such that only a small amount of probe is bound, even at saturation.

3.1.4.1 Comparing Calcium Channel Blockers by Displacement Assay

Agents that block calcium ion channels are of great pharmacological interest for their use in treating hypertension and certain kinds of cardiac arrhythmias. These drugs bind to receptors in cardiac tissue and in vascular tissue. By decreasing the inward current of calcium ions through Ca^{2+} channels, they can slow the conduction of electrical impulses that produce contractions of the heart muscle; this reduces abnormal conduction and heart contractions. Since contraction by vascular smooth muscle cells depends on a flux of calcium ions, these agents also produce vasodilation and help to lower blood pressure. There are three main classes of drugs that act as Ca^{2+} channel blockers, i.e., benzothiazepines, dihydropyridines, and aralkylamines. These three drug classes are structurally quite dissimilar and appear to have distinct receptor sites. Their effect is similar, however, in that they all block the passage of calcium ions through membranes.

The compound *d-cis*-diltiazem is a well-characterized benzothiazepine blocking agent and is readily available with a radioactive tritium label. It can thus serve as a probe in a filtration assay for screening other compounds as displacing agents. Figure 3.2 shows that [^{3}H]-diltiazem binds specifically and tightly to membrane-bound receptors, as required for a suitable probe species. The straight line obtained in the Scatchard-type plot (panel B) is consistent with a single class of binding sites on the membrane. The equilibrium dissociation constant by this assay is 83 nM, for fairly tight binding of the drug to its receptor.

Figure 3.3 summarizes the results of a filtration competition assay in which several different Ca^{2+} channel blockers competed against [^{3}H]-diltiazem. First, the radiolabeled drug can be displaced by *d-cis*-[^{1}H]-diltiazem, with a dissociation constant of 89 nM, which matches that of the tritiated species very nicely. Interestingly, the receptor demonstrates stereoselectivity in drug binding; *l-cis*-diltiazem is much

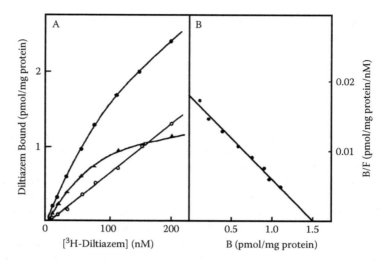

FIGURE 3.2 Binding of diltiazem to purified cardiac sarcolemmal membrane vesicles, using a filtration assay. A. Total binding (●), nonspecific binding (○), and specific binding (▲) as the difference between total and nonspecific binding. B. Scatchard plot of specific binding from A. Adapted with permission from *Journal of Biological Chemistry*, Vol. 261, M.L. Garcia et al., "Binding of Ca^{2+} entry blockers to cardiac sarcolemmal membrane vesicles," pages 8146–8157. Copyright 1986 American Society for Biochemistry and Molecular Biology.

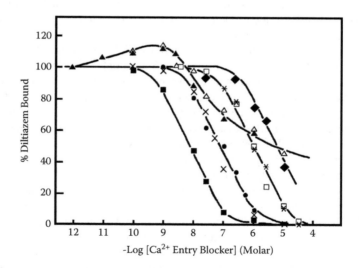

FIGURE 3.3 Effect on [³H]-diltiazem binding to cardiac sarcolemmal membrane vesicles by various Ca^{2+} channel modulators: *d-cis*-diltiazem (●), *l-cis*-diltiazem (♦), (−)-D600 (■), (+)-D600 (□), (±)-verapamil (×), nitrendipine (▲), Bay K 8644 (Δ), and bepridil (*). Adapted with permission from *Journal of Biological Chemistry*, Vol. 261, M.L. Garcia et al., "Binding of Ca^{2+} entry blockers to cardiac sarcolemmal membrane vesicles," pages 8146–8157. Copyright 1986 American Society for Biochemistry and Molecular Biology.

less effective than the *d-cis* isomer in displacing the tritium-labeled drug, with a dissociation constant of only 7 μM.

Verapamil is a representative aralkylamine antagonist of the Ca^{2+} channel, and it and the related compounds D-600 and bepridil compete with $[^3H]$-diltiazem quite effectively. Dissociation constants for verapamil and bepridil are 49 nM and 0.8 nM, respectively.

The compound (−)-D600 has about 10 times higher affinity for the receptor than the probe, with a dissociation constant of 8 nM, and (+)-D600 has about 10 times less affinity than the probe ($k_{diss} = 890$ nM); apparently, stereospecificity in binding extends to the aralkylamines as well.

Finally, the dihydropyrimidines nitrendipine (an antagonist) and Bay K 8644 (an agonist) exhibit complex behavior, stimulating the uptake of diltiazem at low concentrations (note the initial rise in percent diltiazem bound), and then inhibiting its binding at higher concentrations.

3.1.5   COMPETITIVE INHIBITION OF AN ENZYME

The reversible, competitive inhibition of enzymes is a major area in which displacement or competition assays are used routinely. In fact, one of the most commonly used sets of formulas in pharmacology, the Cheng-Prusoff equations, was originally developed to treat enzyme *kinetics* inhibition, and not equilibrium displacement assays [5]. Enzyme assays are typically run with excess substrate, to saturate the enzyme and so get steady-state kinetics, while varying the concentration of inhibitor. This corresponds to performing a set of competitive binding assays by varying one ligand's concentration (it being used at relatively low levels) and adding the other ligand in great excess, so that the free concentration of the latter is scarcely altered over the course of titration.

3.1.5.1   The Cheng-Prusoff Relations

The Cheng-Prusoff equations are based on several assumptions: (1) the two ligand species compete for the same sites; (2) there is only a single class of binding sites present; and (3) the concentrations of both the displaced ligand and the displacing ligand are much greater than the concentration of receptor. (The concentration of free substrate remains effectively equal to its total concentration $[S]_0$, since only a negligibly small fraction is bound.)

A classic competitive inhibitor is a small molecule resembling the substrate, which binds reversibly in the enzyme's active site, and which then blocks reaction with the substrate. The usual representation of this inhibition scheme is

$$E + I \underset{K_I}{\rightleftharpoons} EI$$

$$E + S \underset{K_S}{\rightleftharpoons} ES \xrightarrow{k_{cat}} E + P \tag{3.18}$$

Here K_I and K_S are the dissociation constants for releasing inhibitor I or substrate S from the EI or ES complexes, respectively, while k_{cat} is a rate constant for the conversion of the

ES complex into (released) product P and free enzyme E. The net effect of I is to sequester enzyme as the inactive EI complex and so to reduce the overall rate of reaction.

The standard method of determining these dissociation constants is to start with the Michaelis-Menten equation:

$$v = \frac{k_{cat}\,[E]_t[S]}{K_M + [S]}$$

(3.19)

where v is the velocity or rate of reaction and $[E]_t$ is the total enzyme concentration. This equation is analogous to the Langmuir binding isotherm equation upon identifying $v/(k_{cat}\,[E]_t)$ as the degree of saturation Y, and identifying K_M (effectively a dissociation constant) with $1/k_S$, where k_S is an *association* constant for the uptake of substrate. With a bit of algebra it is easy to show that, for a competitive inhibitor with a dissociation constant K_I, the percent inhibition is given by

$$\% \text{ inhibition} = 100 \times \left(\frac{\left(\dfrac{[I]}{K_I}\right)}{1 + \left(\dfrac{[I]}{K_I}\right) + \left(\dfrac{[S]_0}{K_M}\right)} \right)$$

(3.20)

The concentration that produces 50% inhibition (again, for a competitive inhibitor) can be shown to be

$$IC_{50} = K_I \left(1 + \frac{[S]_0}{K_M} \right)$$

(3.21)

When K_M is much greater that $[S]_0$ (that is, when the substrate concentration is much lower than the K_M value), then it is legitimate to neglect the second term inside the brackets, and the IC_{50} is approximately equal to the K_I value. However, since enzyme kinetic assays are usually done with excess substrate, the IC_{50} will usually be a function of $[S]_0$. To compare IC_{50} values from different laboratories, measurements must be done specifically at the same substrate concentrations, or the comparisons will be meaningless. Similar comments hold for noncompetitive inhibitors as well. As for uncompetitive inhibitors, so long as $[S]_0 \gg K_M$, then IC_{50} will be (approximately) independent of $[S]_0$ and equal to K_I.

The practicality of using IC_{50} values emerges when there are a large number of potential inhibitors to be screened. The time and labor needed to run complete inhibition kinetic studies to determine K_I for each compound make it impractical to screen large numbers of compounds this way. So long as the compounds belong to the same class of inhibitors (that is, all are competitive, or all are uncompetitive, etc.), then it is far easier to run assays designed to evaluate IC_{50} values. One can pick a convenient substrate concentration for the assay, then check the degree of inhibition at just a few concentrations of (potential) inhibitor, enough to deduce the individual IC_{50} values and so rank the candidates. It is then easy to choose the most potent compounds for further study and development.

An important part of the utility of the IC_{50} value is that it can be defined operationally, without reference to a particular kinetic scheme of inhibition. Just as an effective Michaelis constant can be defined for a cooperative enzyme (one that definitely does *not* follow the Michaelis-Menten kinetic scheme) as that value for $[S]_0$ that gives 50% of maximal velocity, so can an IC_{50} value be defined operationally as the inhibitor concentration that gives 50% inhibition (but at a specified value of $[S]_0$, of course). Thus, an IC_{50} value can be used to qualitatively characterize binding to cooperative and multisite systems in which the underlying molecular model may yet be uncertain.

3.1.5.2   Validating the Use of the Cheng-Prusoff Relations

As noted before, the IC_{50} is not a true thermodynamic constant since it depends on the amount of substrate present in the assay mixture. Thus, it is not appropriate to use IC_{50} values in determining thermodynamic quantities like ΔH or ΔC_P. It is necessary to convert the IC_{50} to a K_I first, using Eq. 3.21. This conversion of course assumes the applicability of the Cheng-Prusoff model, so how to check this assumption?

Equation 3.21 shows that the IC_{50} will approach K_I as $[S]_0/K_M$ approaches zero. This suggests plotting both K_I and IC_{50} as functions of the substrate concentration $[S]_0$. If the model is correct, the two plots should converge at low substrate concentrations. Furthermore, the plot of K_I should give an essentially flat line of zero slope, while the plot for IC_{50} should have a positive slope and be everywhere above the flat line for K_I.

Similar considerations will apply to agonist-antagonist binding competition assays. Figure 3.4 (from Ref. 6) shows such a graphical verification of the IC_{50} to K_I conversions, for ligands binding to a β-adrenergic receptor preparation. The lack of dependence of the calculated high-affinity dissociation constant K_H on agonist concentration, together with the convergence of the IC_{50} and K_I plots to a common value, validates the use of the Cheng-Prusoff relations for this system.

3.1.6   FURTHER CONSIDERATIONS IN COMPETITION ASSAYS

The relations discussed here are only approximate, subject to the assumptions listed above. They cannot be used, for example, when the concentration of labeled ligand (species 1) or competing ligand (species 2) is appreciably depleted by binding. Furthermore, at a fundamental level these relations assume that the binding follows a simple mass-action law, ignoring complications from cooperativity, conformational switching, etc.

Under typical enzyme assay conditions, the substrate and the inhibitor will be present in vast excess over the enzyme concentration. However, with inhibitors that bind very tightly, the assays may have to be done with $[I]_t$ comparable to $[E]_t$. Obviously, similar situations may arise with nonenzymatic receptors interacting with agonists or antagonists. Alexander Levitzki and coworkers have analyzed the competition/displacement experiment in detail, taking due account of the concentrations of all relevant species, and have generalized the Cheng-Prusoff equation to allow for arbitrary ratios of ligand/receptor concentrations [7,8].

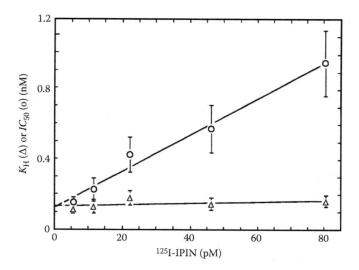

FIGURE 3.4 Validation of the applicability of the Cheng-Prusoff relations for conversion of IC_{50} values to true thermodynamic constants. Dependence of IC_{50} and K_H (inhibitor dissociation constant) values on agonist concentration (radiolabeled iodopindolol, ^{125}I-IPIN), for binding to a β-adrenergic receptor preparation. Adapted with permission from *Journal of Pharmacology and Experimental Therapeutics* Vol. 237, M.L. Contreras, B.B. Wolfe, and P.B. Molinoff, "Thermodynamic properties of agonist interactions with the beta adrenergic receptor-coupled adenylate cyclase system. I. High- and low-affinity states of agonist binding to membrane-bound beta adrenergic receptors," pages 154–164. Copyright 1986 ASPET.

   Henis and Levitzki [9] have examined competitive binding for two common models of cooperative binding to allosteric enzymes, and have shown how to use the slope of a Hill plot at 50% ligand saturation (the Hill coefficient), in the presence and absence of competitor, to decide on a model for the data. Jezewska and Bujalowski [10] have analyzed the case of a single ligand species in the presence of two competing species of macromolecular receptor, where the observed (assumed spectroscopic) binding signal derives from the receptor, not the ligand. Their approach leads to a model-free binding isotherm, which will be considered further in Chapter 6.

3.2   LINKAGE AND "PIGGY-BACK" BINDING

A common biological strategy for regulating enzyme or receptor activity is to use a secondary ligand that, while not binding directly to the receptor itself as the primary ligand does, nevertheless influences the binding of that primary ligand. Perhaps titration of the primary ligand at a particular acidic or basic site is needed in order for it to form a hydrogen bond with a group on the receptor. Or binding of the secondary ligand may drive a conformational change in the primary ligand, which then enables that ligand to make more favorable contacts with the receptor.

This appearance of a secondary ligand is called "piggy-back" binding, and in concept it resembles models involving allosteric effectors, to be discussed in Chapter 4. The difference here, however, is that the "effector" or secondary ligand is acting on the primary *ligand* and *not* directly on the receptor or enzyme. This sort of binding might appear, for example, as the compulsory protonation of an acidic ligand before it may bind to its receptor; see the example of nucleotide inhibitor binding to RNase (Section 3.2.3). Another common situation involves the binding of a metal ion, such as Mg^{2+} or Ca^{2+}, to a protein, which in turn may modulate the binding of this protein to another protein; the regulation of the contraction of skeletal muscle is a good example, in which Ca^{2+} binds to troponin, inducing a conformational change there and providing a signal that leads eventually to myofilament contraction [11]. Piggy-back binding has also been suggested as a model for binding in several different DNA-protein systems, e.g., the binding of cAMP by a gene-regulatory protein, with the complex then binding to regulatory sites on DNA [12].

3.2.1 BASIC THEORY FOR PIGGY-BACK SYSTEMS

In piggy-back binding, a primary ligand species (L_1) binds directly to the receptor, but this primary ligand has on it a binding site for a secondary ligand species (L_2); thus, L_2 binds to L_1 or to the complex of the receptor with L_1, but not to the receptor directly. The complex of L_1 with L_2 forms a third distinguishable species in solution, L_3, so that L_1 and L_3 compete for binding to the receptor. Then for a receptor with a single binding site there are three possible states: unligated, ligated with L_1, and ligated with the complex species L_3.

One way to write the binding polynomial for such a system would simply be

$$\mathbf{P} = 1 + k_1[L_1] + k_3[L_3] \tag{3.22}$$

where k_1 and k_3 are the binding constants for the uncomplexed ligand species L_1 and for the complex L_3 species, respectively. Formally, this represents L_1 and L_3 as competitors for the same binding site, with binding densities $< r_1 >$ and $< r_3 >$, respectively, and an overall binding density given by

$$< r > = < r_1 > + < r_3 > = \frac{k_1[L_1]}{\mathbf{P}} + \frac{k_3[L_3]}{\mathbf{P}} \tag{3.23}$$

The binding density is a function of two concentrations, $[L_1]$ and $[L_3]$, as would be expected for a competitive binding situation.

This formulation would be adequate if it were possible to monitor both L_1 and L_3 simultaneously throughout the titration. But instead suppose that the experimental method gives no direct measure of species 3, but only information on $[L_1]$ and $[L_2]$ (say, absorbance for L_1, and fluorescence or ionic activity for L_2). There is, however, a connection among the three species: $[L_3] = [L_1] [L_2] k_{1-2}$, where k_{1-2} governs the equilibrium between L_1 and L_2 to form L_3 in solution. Figure 3.5 illustrates the various equilibria involved.

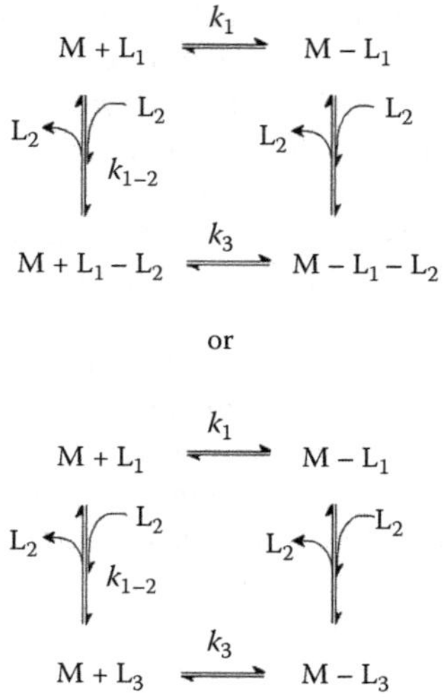

FIGURE 3.5 A simple example of piggy-back binding. The receptor M binds ligand L_1; L_1 can bind a second species L_2 to form the complex L_3, which can then be taken up by the receptor.

Substitution for $[L_3]$ in the expression for **P** and factorization of the result gives

$$\mathbf{P} = 1 + k_1[L_1]\left(1 + \frac{k_{1-2}k_3}{k_1}[L_2]\right) \tag{3.24}$$

The binding density is then

$$<r> = \frac{k_1[L_1]\left(1 + \frac{k_{1-2}k_3}{k_1}[L_2]\right)}{\mathbf{P}} \tag{3.25}$$

Note that the binding density is still a function of two species, L_1 and now L_2. Variations in the concentration of L_2 will affect the binding density overall and of course the total amount of L_1 that is taken up by the macromolecule. If no secondary ligand L_2 is present, then the result is a simple Langmuir isotherm equation, with $<r> = k_1[L_1]/(1 + k_1[L_1])$.

Notice that adding the secondary ligand L_2 may drive further binding of the primary ligand species L_1 onto the receptor if the L_1-L_2 complex has a higher affinity for the receptor than does the simple primary ligand L_1. Conversely, if the L_1-L_2 complex has a lower affinity than L_1, then there may be a net decrease in the overall binding to the receptor; in this case the L_1-L_2 complex serves to deplete the solution of the higher affinity ligand L_1.

Certain cases may present the extreme of piggy-back binding, in which *only* the species L_3 can bind. Here the binding polynomial is

$$P = 1 + K[L_1][L_2] \tag{3.26}$$

The lumped constant K accounts for both binding to the macromolecule and the preceding formation of the L_1-L_2 complex. The equation for the binding density becomes

$$<r> = \frac{K[L_1][L_2]}{1 + K[L_1][L_2]} \tag{3.27}$$

which in terms of the piggy-back species L_3 is simply

$$<r> = \frac{k_3[L_3]}{1 + k_3[L_3]} \tag{3.28}$$

An important instance of piggy-back binding that involves a linear lattice of sites, e.g., a gene-regulatory protein binding to DNA, will be discussed in Chapter 5.

3.2.2 HYDROGEN ION AS A PIGGY-BACK LIGAND: THEORY FOR pH EFFECTS

The proton, H^+, is the simplest and probably the most common biochemical ligand. Its effects are ubiquitous: a change in pH can shift conformational equilibria, perturb reaction kinetics, reverse or promote protein aggregation, etc. Since the protons bind to specific titrable residues and have distinct affinities, it is usually straightforward to set up binding polynomials for the reactants and products. The analysis of pH effects on the overall free energy change for the observed equilibrium can be more difficult, however.

3.2.2.1 Titration of a Single Residue on the Ligand

All that is needed for this simple model is a short revision of the original piggy-back formula for P. Simply replace L_2 with H^+, and replace the ratio (k_3/k_1) with K_a^0/K_a', a ratio of the acidity (dissociation) constant for the residue on the free ligand to that for the receptor-bound ligand:

$$P = 1 + k_1[L_1]\left(1 + k_{1-H^+} \frac{K_a^0}{K_a'}[H^+]\right) \tag{3.29}$$

(This expression follows common practice in using *dissociation* constants K_a' for pH effects, but retains our convention in using k_1 and k_{1-H^+} for association constants.)

The binding density can be expressed as an explicit function of [H⁺] and the acid dissociation constants of the titrable residue on the ligand:

$$< r_1 > = \frac{k_1[L_1]\left(1 + k_{1-H^+}\dfrac{K_a^0}{K_a{'}}[H^+]\right)}{P} \tag{3.30}$$

or, collecting terms and hiding the dependence on [H⁺] in an effective binding constant for L_1,

$$< r_1 > = \frac{k_{eff}[L_1]}{1 + k_{eff}[L_1]}, \qquad k_{eff} = k_1\left(1 + k_{1-H^+}\frac{K_a^0}{K_a{'}}[H^+]\right) \tag{3.31}$$

This formalism could be applied, for example, to the acid titration of a receptor agonist or antagonist as it binds to receptor, or to the titration of an enzyme inhibitor complexing with enzyme.

A pertinent example here is the binding of the antifolate agent methotrexate (MTX) to the enzyme dihydrofolate reductase from various bacteria and fungi (e.g., *Streptococcus faecium*, *Lactobacillus casei*, and *Escherichia coli*) [13–15]. Strong binding of the inhibitor MTX to this enzyme seems to depend on protonation at N − 1 on the pteridine ring of methotrexate (Figure 3.6), with subsequent ion pair formation of this moiety with a carboxylate residue in the active site of the enzyme. This interaction appears to be a key to achieving high binding affinity; crystal structures

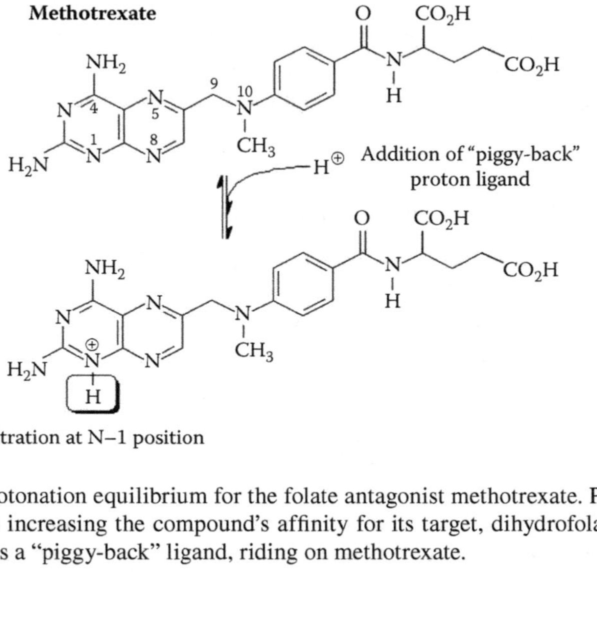

FIGURE 3.6 Protonation equilibrium for the folate antagonist methotrexate. Protonation at N − 1 is a key to increasing the compound's affinity for its target, dihydrofolate reductase. The proton here is a "piggy-back" ligand, riding on methotrexate.

of DHFR with trimethoprim (another folate analog) show a similar ionic association of the drug's protonated pyrimidine ring with the enzyme's carboxylate moiety.

3.2.2.2 Comparison to Titration of a Single Residue on the Receptor

For a single receptor macromolecule, with separate binding sites for L_1 and H^+, there are four possible states: $M^{(0)}$ (unligated), ML, MH, and MLH. The binding polynomial is therefore

$$\mathbf{P} = 1 + k_1[L_1] + \frac{[H^+]}{K_a^0} + \alpha k_1[L_1]\frac{[H^+]}{K_a^0} \tag{3.32}$$

where K_a^0 is now the acid dissociation constant for the site on the *macromolecule*, and α is a modifying factor (a *cooperative interaction factor*; more on this in Chapter 4) to allow for interaction between the two occupied sites. The binding density with respect to L_1 is

$$<r_1> = \frac{k_1[L_1]\left(1 + \alpha\frac{[H^+]}{K_a^0}\right)}{\mathbf{P}} \tag{3.33}$$

which can be written as

$$<r_1> = \frac{k_{eff}[L_1]}{1 + k_{eff}[L_1] + \frac{[H^+]}{K_a^0}}, \qquad k_{eff} = k_1\left(1 + \alpha\frac{[H^+]}{K_a^0}\right) \tag{3.34}$$

which parallels the previous result for titration of the ligand. This formulation could also be applied to the treatment of the proton as an allosteric effector that modulates the receptor directly.

The general approach here is easily extended to accommodate multiple sites of protonation, both on the ligand and on the macromolecule, and it can include cooperative effects as well.

3.2.3 pH Effects in RNase-Inhibitor Binding

The pH dependence of bovine pancreatic ribonuclease A (RNase A) binding the inhibitory mononucleotide 3′-cytidine monophosphate (3′-CMP) has been quantitatively characterized by spectrophotometric assays and by calorimetry [16,17]. Nucleotide binding to the enzyme follows a 1:1 stoichiometry, and it appears that the pH dependence stems from titration of three groups (histidine residues) on the enzyme, together with the phosphate group on the nucleotide. This is a moderately complicated binding system, involving proton binding to both the macromolecule and to the primary ligand, 3′-CMP. The microscopic model presented here will show how to handle both kinds of binding at the same time.

The simplest scheme to explain the pH dependence is to suppose that all proton binding sites are acting independently of one another, and that over the pH range from 4 to 9 only the dianionic phosphate form of the nucleotide is bound to the enzyme. Since the acidity of a histidine residue depends on its immediate chemical environment, the binding of the nucleotide onto the protein could quite possibly cause a change or shift in the pK_a values of the histidines. Thus, it is necessary to account for titration of the three histidines, first in the uncomplexed RNase A molecule, and second in its complex with 3′-CMP. An expression for the observed equilibrium binding constant that takes all this into account would be

$$\ln K_{obs} = \ln K_{th} + \sum_{j=1}^{3} \ln\left(1 + k_j' \,[\text{H}^+]\right) - \sum_{j=1}^{3} \ln\left(1 + k_j[\text{H}^+]\right) - \ln\left(1 + k_P[\text{H}^+]\right) \quad (3.35)$$

Here k_j and k_j' refer to the proton binding constants of the j-th histidine residues of free RNase and of the complex, respectively, while k_P refers to the proton binding constant of the phosphate moiety of the nucleotide whose proton affinity does not change. Notice that the bound form of the nucleotide is assumed to be the dianionic nucleotide, so there is no term involving a k_P'. For comparison, the relation can be recast in terms of acid dissociation constants K_j and K_j':

$$\ln K_{obs} = \ln K_{th} + \sum_{j=1}^{3} \ln(1 + [\text{H}^+]/K_j') - \sum_{j=1}^{3} \ln(1 + [\text{H}^+]/K_j) - \ln(1 + [\text{H}^+]/K_P)$$

$$(3.36)$$

Flogel and Biltonen [17] found that a quantitative fit of their calorimetric data could be obtained with $pK_1 = 5.0$, $pK_2 = 5.8$, $pK_3 = 6.7$, $pK_1' = 6.7$, $pK_2' = 7.3$, $pK_3' = 7.0$, and $pK_P = 6.1$, with $\Delta G_{th}° = -14.3$ kJ/mol. The nucleotide binding-induced shifts in acidity of the histidine residues seem reasonable.

Figure 3.7, taken from Flogel and Biltonen [17], shows how the (negative of the) free energy change for binding 3′-CMP with RNase varies with pH; panel B includes a correction for the ionization of the phosphate group of 3′-CMP (it can be seen that this correction is quite substantial at acid pH). The model of multiple titrable residues with different proton affinities leads to the strongly curved theoretical plot shown. The model appears to account quite successfully for the observed pH dependence of nucleotide binding by RNase A.

3.3 LINKAGE EFFECTS ON MACROMOLECULAR ASSOCIATIONS AND CONFORMATIONAL CHANGES

The connection of ligand binding to conformational changes in macromolecules can be a delicate matter; changes in the concentration of a ligand species may trigger large and consequential changes in a macromolecule or macromolecular assembly. Such effects are highly important for the stability of proteins and nucleic acids in solution:

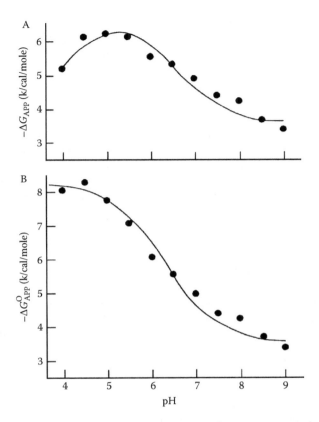

FIGURE 3.7 Influence of pH on the binding of 3'-CMP by RNase A. Adapted with permission from *Biochemistry* Vol. 14, M. Flogel and R.L. Biltonen, "The pH dependence of the thermodynamics of the interaction of 3'-cytidine monophosphate with ribonuclease A," pages 2610–2615. Copyright 1975 American Chemical Society.

ligands may selectively stabilize or destabilize particular macromolecular conformations, and cause denaturation, precipitation, solubilization, aggregation, etc. Ligand binding is also found to be linked to the association/dissociation of macromolecular complexes such as multisubunit enzymes, cytoskeletal scaffolding assemblies, DNA-protein complexes, and many other macromolecular assemblies. The thermodynamics of how these processes are linked together, *linkage thermodynamics* or simply *linkage*, for short, is at the base of a host of biological processes.

3.3.1 Linkage and an A $\rightleftharpoons$ B Equilibrium

Consider a macromolecule that can switch between two different states, A and B, both of whose concentrations can be observed simultaneously. The two-state equilibrium here could describe a conformational change in the macromolecule, two different orientations of the macromolecule with respect to some external field, etc.

For the observed equilibrium (that is, for the *observed concentrations* of A and B, not necessarily their thermodynamic activities), the observed equilibrium constant K_{obs} and the observed free energy change ΔG^0_{obs} are

$$A \rightleftharpoons B, \quad K_{obs} = \frac{[B]}{[A]}, \quad \Delta G^0_{obs} = -RT \ln K_{obs} \tag{3.37}$$

Now suppose that a third solution component, the ligand L, can bind to either or both of the species A and B. The species L might be a cofactor for an enzyme, a proton, a small inorganic ion, or a denaturing agent like urea or guanidinium. Changes in the concentration (more properly, the thermodynamic activity) of L will now shift the equilibrium and produce an apparently different equilibrium constant. The aim is to deduce how K_{obs} varies with changes in the activity of L.

For generality, suppose there is some degree of uptake of L by both A and B. The participation of L can be shown more explicitly:

$$A \bullet L_{<r_{L,A}>} \rightleftharpoons B \bullet L_{<r_{L,A}>} + \Delta r_L L \tag{3.38}$$

Here A•L and B•L represent the complexes of L with A and B, respectively; the subscripted quantities $< r_{L,A} >$ and $< r_{L,B} >$ indicate the amounts of L bound to A and to B, per mole respectively, and Δr_L is defined as the difference in binding density of L across the equilibrium:

$$\Delta r_L = < r_{L,A} > - < r_{L,B} > \neq 0 \tag{3.39}$$

As written, the equilibrium shows an apparent release of ligand, but note that Δr_L could be either positive or negative; positive if there is net release in passing from A to B, but negative if there is net uptake of L in passing from A to B.

Now suppose the original observed equilibrium constant is amended to include the species L, and thermodynamic activities are introduced in place of simple concentrations. The result is the *true thermodynamic* equilibrium constant for this system:

$$K_{th} = \frac{a_B}{a_A} a_L^{\Delta r_L} \tag{3.40}$$

The associated true thermodynamic free energy change, $\Delta G^0_{th} = -RT \ln K_{th}$, is connected to the observed free energy change through

$$\Delta G^0_{obs} = \Delta G^0_{th} + RT \ln \left(\frac{\gamma_B}{\gamma_A} \right) + \Delta r_L \left(RT \ln a_L \right) \tag{3.41}$$

The macromolecular activity coefficients γ_A and γ_B may be taken as equal to each other, provided that the ligand L does not appreciably alter the charge, size, or shape of the macromolecule when it binds (that is, there is no appreciable change in the *thermodynamic activity* of the macromolecule). Under these conditions it is reasonable to ignore the second term on the right-hand side of the equation.

Now take the derivative of this last expression with respect to $\ln a_L$. Note that K_{th} is a true constant, so its derivative is zero; the derivative of the term in γ_A and γ_B is (approximately) zero as well, provided that the ligand has no appreciable effect on the activity of the macromolecule. Replacing $\Delta G°_{obs}$ with $-RT \ln K_{obs}$ will give

$$\left(\frac{\partial \ln K_{obs}}{\partial \ln a_L} \right)_{T,P} = -\Delta r_L \qquad (3.42)$$

This relation suggests a route to values for Δr_L: measure K_{obs} as a function of a_L, and construct a double-logarithmic plot of $\ln K_{obs}$ versus $\ln a_L$. The result should be a straight line with a slope equal to $-\Delta r_L$. If the ligand L binds only to the B form, say, then $-\Delta r_L$ corresponds to the net uptake of L by B; that is, the slope of the plot equals $< r_{L,B} >$ when $< r_{L,A} > = 0$. If the binding of L is tight and follows a mass-action law, such a double-logarithmic plot can reveal the binding stoichiometry of L on B.

3.3.2 LIGAND-LIGAND LINKAGE AND AN $A + B \rightleftharpoons C$ EQUILIBRIUM

Now consider an apparently simple equilibrium of A and B forming the equilibrium complex C (two proteins associating, for example, or a protein binding to a nucleic acid):

$$A + B \rightleftharpoons C \qquad (3.43)$$

Assume now that the true equilibrium also involves the uptake or release of two ligand species, L_1 and L_2. The (hypothesized) true equilibrium can be written as

$$A \bullet (L_1)_{<r_1,A>} \bullet (L_2)_{<r_2,A>} + B \bullet (L_1)_{<r_1,B>} \bullet (L_2)_{<r_2,B>} \rightleftharpoons C \bullet (L_1)_{<r_1,C>} \bullet (L_2)_{<r_2,C>}$$
$$+ \Delta r_1 L_1 + \Delta r_2 L_2 \qquad (3.44)$$

where $A \bullet (L_1) \bullet (L_2)$, $B \bullet (L_1) \bullet (L_2)$, and $C \bullet (L_1) \bullet (L_2)$ are the ligand complexes with species A, B, and C. The bracketed subscripts on the parenthesized quantities indicate the stoichiometry of the respective complexes. The changes in binding density of the two ligand species across the equilibrium are

$$\Delta r_1 = < r_{1,A} > + < r_{1,B} > - < r_{1,C} > \neq 0 \qquad (3.45)$$

and

$$\Delta r_2 = < r_{2,A} > + < r_{2,B} > - < r_{2,C} > \neq 0 \qquad (3.46)$$

Next, write out the *true* thermodynamic equilibrium constant, using thermodynamic activities instead of simple concentrations, to include all five participating species:

$$K_{th} = \left(\frac{a_C}{a_A a_B} \right) a_1^{\Delta r_1} a_2^{\Delta r_2} = K_{obs} \left(\frac{\gamma_C}{\gamma_A \gamma_B} \right) a_1^{\Delta r_1} a_2^{\Delta r_2} \tag{3.47}$$

The true thermodynamic free energy change is of course $\Delta G_{th}^0 = -RT \ln K_{th}$. Now, the connection between observed and true thermodynamic free energy change is

$$\Delta G_{obs}^0 = \Delta G_{th}^0 + RT \ln \left(\frac{\gamma_C}{\gamma_A \gamma_B} \right) + \Delta r_1 \left(RT \ln a_1 \right) + \Delta r_2 \left(RT \ln a_2 \right) \tag{3.48}$$

Then, using the expressions for equilibrium constants K_{obs} and K_{th}, noting that K_{th} is a true constant, and taking the derivative with respect to the activity of species L_1, gives

$$\left(\frac{\partial \ln K_{obs}}{\partial \ln a_1} \right)_{T,P} = -\Delta r_1 - \Delta r_2 \left(\frac{\partial \ln a_2}{\partial \ln a_1} \right)_{T,P} \tag{3.49}$$

Under many solution conditions the last term on the right may be negligibly small, the ligand L_1 having little or no effect on the activity of ligand L_2 and vice versa, so that the derivative $\partial \ln a_2 / \partial \ln a_1$, even multiplied by Δr_2, is much smaller than Δr_1. Under these conditions a plot of the logarithm of K_{obs} as a function of the logarithm of the ligand activity should be (approximately) a straight line with a slope equal to Δr_1. That is, this double-logarithmic plot can give the net uptake or release of ligand across the equilibrium.

Notice, however, that if the activity of L_2 is influenced by ligand L_1 to even a moderate degree, then it is necessary to carefully consider the magnitude of the last term on the right. For example, if ligand L_1 is present in high concentrations (as with urea at 6 M concentration, to dissociate protein subunits, say), then osmotic effects can raise the activity of water (here, species L_2) past its simple molar concentration. The combination of the derivative $\partial \ln a_2 / \partial \ln a_1$ with even a modest exchange of waters Δr_2 might approach the magnitude of Δr_1 and so confound a simple interpretation of the double-logarithmic plot. Curvature in the plot of $\ln K_{obs}$ as a function of $\ln a_1$ would be a tip-off to this kind of situation.

3.3.3 General Expression for the Salt Dependence of K_{OBS}

To include the linkage of salt with ligands and their multifarious effects on a macromolecular equilibrium $A + B \rightleftharpoons C$, one approach is to develop and apply a formal generalized binding polynomial, as shown by Record and coworkers [18]. This approach allows for multiple ligand species, including solvent water (W), ions (cation M and anion X), and other small ligand species (L). The basic scheme is to formulate the observed equilibrium constant K_{obs} as a phenomenological equilibrium quotient K^0 for the $A + B \rightleftharpoons C$ equilibrium, supplemented with binding polynomials for water and ligand:

$$K_{obs} = K^0 \times \left(\frac{\mathbf{P}_C^0}{\mathbf{P}_A^0 \, \mathbf{P}_B^0} \right) \qquad (3.50)$$

The completely unligated condition with respect to L, W, M, and X could be chosen as the thermodynamic reference state, but this is not strictly necessary, and in fact such a state may not be experimentally accessible (complete dehydration of many biopolymers leads to loss of their activities, so binding to a biopolymer in such a state is hypothetical at best). The main interest in these cases is usually in comparing *relative* gains or losses of L, W, M, or X. To do this, some other, more accessible state can be chosen as the reference state. The single superscript zero will indicate this reference state.

To examine the effect of, say, added salt on the observed binding, Record et al. [18] considered how K^0 depends on activity of the macromolecular species observed (A, B, and C). The explicit form of K^0 is

$$K^0 = K_{th}^0 \left(\frac{\gamma_A^0 \gamma_B^0}{\gamma_C^0} \right) \qquad (3.51)$$

They could then write the observed equilibrium constant

$$K_{obs} = K_{th}^0 \times \left(\frac{\gamma_A^0 \, \gamma_B^0}{\gamma_C^0} \right) \times \left(\frac{\mathbf{P}_C^0}{\mathbf{P}_A^0 \, \mathbf{P}_B^0} \right) \qquad (3.52)$$

The corresponding free energy change could be written as

$$\Delta G_{obs} = \Delta G_{th}^0 - RT \ln P_C^0 + RT \ln P_A^0 + RT \ln P_B^0 - RT \ln \left(\frac{\gamma_A^0 \, \gamma_B^0}{\gamma_C^0} \right) \qquad (3.53)$$

and the differential $d \ln K_{obs}$ is

$$d \ln K_{obs} = d \ln \mathbf{P}_C^0 - d \ln \mathbf{P}_A^0 - d \ln \mathbf{P}_B^0 + d \ln \left(\frac{\gamma_A^0 \gamma_B^0}{\gamma_C^0} \right) \qquad (3.54)$$

As an aside, by taking appropriate derivatives of these expressions, it is possible to go on to derive formulas for the enthalpy and entropy changes, the heat capacity change, etc., and relate them to solvent and solute compositional changes. Such relations can be applied in fitting theoretical models (binding polynomials) to experimental data (ΔH, ΔS, ΔC_P, etc.). However, a more immediate application is in how $\ln K_{obs}$ will change as the activities of the solutes L, W, M, and X may change, that is, how the equilibrium will shift in response to changes in concentration of these ligands, which now may include the solvent and small salt ions.

Focusing on the salt dependence of K_{obs}, Record et al. considered the particular case in which M and X are ionic species, with the salt MX at a molal concentration m (note the choice of concentration units here; this is typical of physical chemistry investigations, though quite *untypical* of biochemical or pharmacological ones) [18]. The mean ionic activity of this solute is

$$a_{\pm} = a_{MX}^{1/p} = (a_M^{p_+} \, a_X^{p_-})^{1/p} \tag{3.55}$$

with the stoichiometry parameter $p = p_+ + p_-$; for a 1:1 salt $p_+ = p_- = 1$ and $p = 2$. They assumed that the dependence of the activity of the solvent water on salt activity could be expressed as

$$d \ln a_W = -\left(\frac{m \, p}{55.6}\right) d \ln a_{\pm} \tag{3.56}$$

which is an approximation that is good for dilute solution (the factor of 55.6 accounts for the molality of pure water). They were then able to arrive at the following general expression for the variation of K_{obs} with ionic activity:

$$\left(\frac{\partial \ln K_{obs}}{\partial \ln a_{\pm}}\right)_{T,P} = -\Delta\left(r_M + r_X - \frac{pm}{55.6} r_W\right)_{T,P} + \left(\frac{\partial \ln[\gamma_A \gamma_B/\gamma_C]}{\partial \ln a_{\pm}}\right)_{T,P} - \Delta r_L \left(\frac{\partial \ln a_L}{\partial \ln a_{\pm}}\right)_{T,P}$$

$$\tag{3.57}$$

This relation can be applied to all manner of macromolecular equilibria, including solubility and phase partitioning, conformational changes, and binding. It includes five potential sources of an effect of an electrolyte MX on the observed equilibrium:

1. Differential cation binding ($\Delta r_M \neq 0$)
2. Differential anion binding ($\Delta r_X \neq 0$)
3. Differential hydration ($\Delta r_W \neq 0$)
4. Differential screening (of the Debye-Hückel type) of the different macromolecular species (an effect of $a_{\pm}$ on one or more of the activity coefficients γ_A, γ_B, or γ_C)
5. Effects of mean ionic activity $a_{\pm}$ on the activity of the ligand L

These may act singly or in combination, depending on the particular macromolecular system and its equilibria, the ligand chosen, and the electrolyte used. The review [18] details several applications of Eq. 3.57, one of which (the dimerization of α-chymotrypsin) is discussed below.

3.3.4 APPLICATIONS OF LOG-LOG PLOTS

Returning to the binding polynomials P_A, P_B, and P_C, suppose the situation is simplified to the binding of a single ligand L (whose concentration is very low) that perturbs the equilibrium $A + B \rightleftharpoons C$. For reference, the binding polynomials here are

$$\mathbf{P_A} = (1 + k_A a_L)^{n_A}, \quad \mathbf{P_B} = (1 + k_B a_L)^{n_B}, \quad \mathbf{P_C} = (1 + k_C a_L)^{n_C} \tag{3.58}$$

It is convenient now to choose the unligated form of the macromolecule as the reference state; now the binding polynomials will reflect a number n_A of equal and independent sites on A with affinity k_A, likewise, a number n_B of equal and independent sites on B with affinity k_B, and finally, a number n_C of equal and independent sites on C with affinity k_C. (This tacitly assumes that binding for all species has well-defined stoichiometries and affinities. This assumption may not apply if the binding is very weak, when the binding may not obey a stoichiometric mass-action equation.)

It is convenient also to approximate the ratio $\gamma_A \gamma_B / \gamma_C$ by unity (these species are all dilute, and so should behave ideally). Then the observed binding constant is

$$\ln K_{obs} = \ln K_{th} + n_C \ln(1 + k_C a_L) - n_A \ln(1 + k_A a_L) - n_B \ln(1 + k_B a_L) \tag{3.59}$$

and its dependence on ligand activity is

$$\left(\frac{\partial \ln K_{obs}}{\partial \ln a_L} \right)_{T,P} = n_C Y_C - n_A Y_A - n_B Y_B = -\Delta r_L \tag{3.60}$$

Similar relations can easily be obtained with different binding models and the corresponding binding polynomials. They all suggest that the net uptake or release of ligand (Δr_L) can be determined from the slope in a plot of $\ln K_{obs}$ as a function of $\ln a_L$.

Note that K_{th} does not appear here, since it is a true constant, independent of the activity of the species L. Note also the minus sign here, which must be duly taken into account when predicting the direction of shifts in the apparent equilibrium as the ligand's concentration (or more properly, its activity) changes. For example, with net release of L, this formula predicts that with an increase in L's activity (by raising its concentration, for example) there will be an apparent shift in the equilibrium toward the reactant side. This, of course, is just what would be predicted using Le Chatelier's principle and the true equilibrium representation.

The log-log plot is a very useful device when dealing with linkage phenomena. For example, it can be used to check on the salt or pH sensitivity of the (apparent) binding constant; if H^+ or salt ions participate in the binding equilibrium, then there should be a nonzero slope in a plot of $\log K_{app}$ versus the pH or the logarithm of the salt concentration. However, the derivative $\partial \ln K_{obs} / \partial \ln a_L$ is generally a function of a_L, and so (even though activities are used, and not simple concentrations) the plot should not be expected to be linear over any extended range in concentration for the ligand. See Equation 3.49 and the discussion that follows it.

3.3.4.1 Salt Effects in α-Chymotrypsin Dimerization

The salt-sensitive dimerization of α-chymotrypsin presents an opportunity to apply the derivative relations connecting K_{obs} with activities of various solutes. Aune et al. [19] and

Aune and Timasheff [20] have studied both salt and pH effects on the protomer (P)-dimer (D) equilibrium for this protein:

$$2P \rightleftharpoons D, \quad K_{obs} = \frac{[D]}{[P]^2} \tag{3.61}$$

The addition of salt (NaCl or $CaCl_2$) promotes the dimerization. Figure 3.8 (from Aune et al. [19]) displays the dependence of $\ln K_{obs}$ on the logarithm of the activity of added NaCl or $CaCl_2$ at a constant pH of 4.1.

The plots for both electrolytes are linear, with very little if any curvature. Although Aune et al. proposed a role for differential hydration and ionic screening in the dimerization, this was later disputed by Record and coworkers in a reanalysis of the data [18], who found that the data could be fit simply with the equations

$$\ln K_{obs} = 14.2 + (1.06 \pm 0.05) \ln a_{NaCl}$$
$$\ln K_{obs} = 14.1 + (0.5 \pm 0.1) \ln a_{CaCl_2} \tag{3.62}$$

Using the relations $a_{NaCl} = a^2_\pm$ and $a_{CaCl_2} = a^3_\pm$, Record et al. [18] found that the derivative $(\partial \ln K_{obs}/\partial \ln a_\pm)$ was 2.1 ± 0.1 in NaCl, and 1.5 ± 0.3 in $CaCl_2$. Since the experiments were done at a pH substantially below the isoelectric point of the protein (pI $\approx$ 6.6), the binding species was likely the chloride anion, a conclusion supported by earlier work [21] on ionic binding to chymotrypsin. Because both log-log plots were linear over such a wide range of salt concentrations, Record et al. concluded

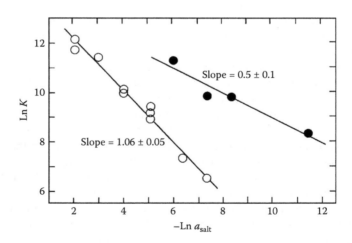

FIGURE 3.8 Salt effects on the dimerization of α-chymotrypsin. Dependence of the dimerization equilibrium constant on ionic activity. NaCl plus 0.01 M acetate buffer pH 4.1 (o); $CaCl_2$ plus 0.01 M acetate buffer pH 4.0 (●). Adapted with permission from *Biochemistry* Vol. 10, K.C. Aune, L.C. Goldsmith, and S.N. Timasheff, "Dimerization of α-chymotrypsin. II. Ionic strength and temperature dependence," pages 1617–1622. Copyright 1971 American Chemical Society.

that ions participated directly and stoichiometrically in the dimerization reaction, with the uptake of approximately one chloride ion per protomer incorporated into dimer, and that there were negligible effects of differential hydration or of Debye-Hückel screening.

3.3.4.2   Water Activity and Chloride Ion Binding in the Oxygenation of Hemoglobin

Chloride ion is known to bind to hemoglobin, and it has been proposed as a possible allosteric regulator of the conformational transition of this protein upon binding oxygen. It has been thought that upon oxygenation, hemoglobin releases 1.6 chloride ions [22]. However, it has been difficult to reconcile this apparent nonintegral stoichiometry with distinct binding sites for the ions or with simple models of the binding. The issue was reexamined by Colombo et al. [23]. Earlier work by these investigators showed that there is a substantial change in hydration of the hemoglobin tetramer upon oxygenation, interpreted as the uptake of about 65 water molecules [24]. In their 1994 study, Colombo et al. controlled the activity of the water by addition of NaCl and sucrose, unlike earlier studies that did not fix the solvent activity [23]. When the binding of chloride ion was studied with fixed water activity (Figure 3.9), the stoichiometry was found to be release of 1.0 chloride, an integral

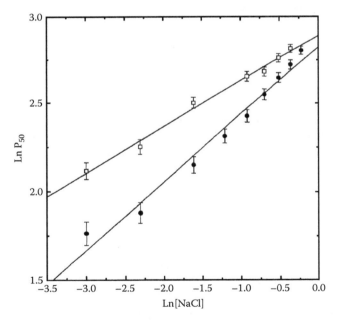

FIGURE 3.9 Dependence on salt concentration of oxygen partial pressure at half saturation (P_{50}) for hemoglobin oxygenation, showing the effects of controlling the thermodynamic activity of water. Without controlling water activity (○); water activity fixed at 1.50 ± 0.05 osmoles/kg of water (●). Adapted with permission from *Proceedings of the National Academy of Sciences USA* Vol. 91, M.F. Colombo, D.C. Rau, and V.A. Parsegian, "Reevaluation of chloride's regulation of hemoglobin oxygen uptake: The neglected contribution of protein hydration in allosterism," pages 10517–10520. Copyright 1994 National Academy of Sciences, U.S.A.

number that accords much better with structural models of the hemoglobin tetramer as well as with chemical intuition. The earlier observation of a release of 1.6 ions per tetramer could now be explained in terms of the dependence of the activity of the solvent water on the chloride ion concentration. The thermodynamic linkage of chloride release and water uptake upon oxygenation of the protein was thus neatly demonstrated by the manipulation of water activity.

3.3.5 UPTAKE OF L_2 BY A MACROMOLECULE PARTIALLY SATURATED WITH L_1

In real biochemical systems it is almost never feasible to work with the completely unligated macromolecular species. Ions will dissociate, acidic or basic moieties will be titrated to various extents, and even the degree of hydration can vary, especially when dealing with very concentrated solute solutions. Thus, it becomes necessary to work out how to treat (at least at a formal level) the uptake of ligand species L_2 when species L_1 is present and some of L_1 is already bound to the macromolecule. The question is, how many molecules of species L_1 are exchanged—released or taken up—when it is known that (on average) j molecules of species L_2 have been bound per macromolecule?

Consider a system with two ligand species L_1 and L_2 that bind to the macromolecule M. The binding polynomial for this system is

$$\mathbf{P} = \sum_{i=0}^{n} \sum_{j=0}^{m} K^{(i,j)} a_1^i a_2^j \tag{3.63}$$

where thermodynamic activities have been used in place of simple concentrations for the ligands. For compactness and clarity, the parenthesized superscripts on K and M indicate the state of ligation. For example, the quantity $K^{(i,j)}$ describes the simultaneous uptake of i molecules of L_1 and j molecules of L_2:

$$M^{(0,0)} + i\,L_1 + j\,L_2 \rightleftharpoons M^{(i,j)}$$

$$K^{(i,j)} = \frac{[M^{(i,j)}]}{[M^{(0,0)}]} a_1^i a_2^j \tag{3.64}$$

Here $M^{(0,0)}$ denotes a macromolecule without any bound ligands of either species, while $M^{(i,j)}$ indicates a macromolecule with i molecules of ligand 1 bound and j molecules of ligand 2 bound.

Alternatively, consider the binding equilibria for species L_2 in the absence of any L_1 (that is, fixing $i = 0$):

$$M^{(0,0)} + j\,L_2 \rightleftharpoons M^{(0,j)}$$

$$K^{(0,j)} = \frac{[M^{(0,j)}]}{[M^{(0,0)}]} a_2^j \tag{3.65}$$

Most importantly, there is also the binding of L_2 in the presence of L_1, where it is necessary to account for all subspecies of macromolecules that have $i = 0, 1, 2, \ldots$ molecules of L_1 bound:

$$M^{(\bullet,0)} + j\,L_2 \rightleftharpoons M^{(\bullet,j)}$$

$$K^{(\bullet,j)} = \frac{[M^{(\bullet,j)}]}{[M^{(0,0)}]}\,a_2^j \tag{3.66}$$

The quantity $[M^{(\bullet,0)}]$ is the total concentration of macromolecules with no L_2 bound, allowing for all states of ligation with L_1.

The equilibrium just described is the one that would probably be followed in most experiments, that is, the binding of L_2 to a macromolecule already in equilibrium with L_1, and not the uptake of L_2 by a totally unligated macromolecule. To analyze such equilibria, it would first be helpful to have an expression for $K^{(\bullet,j)}$ in terms of binding polynomials for the uptake of L_1 by macromolecules with either zero or j molecules of L_2 bound. From that it is possible to show how the derivative of $K^{(\bullet,j)}$ with respect to a_1 is related to the number of molecules of L_1 that are exchanged, per macromolecule, when j molecules of L_2 are taken up.

Suppose there are j molecules of L_2 bound by the macromolecule, but no L_1 molecules. The uptake of L_1 in this case is

$$M^{(0,j)} + i\,L_1 \rightleftharpoons M^{(i,j)}$$

$$K'^{\,(i,j)} = \frac{[M^{(i,j)}]}{[M^{(0,j)}]}\,a_1^i \tag{3.67}$$

(The prime here serves to distinguish this equilibrium constant from that for the simultaneous uptake of both L_1 and L_2.) The total concentration of macromolecules with j molecules of L_2 bound, regardless of how many L_1 are bound, is a sum of the various concentrations $[M^{(i,j)}]$:

$$[M^{(\bullet,j)}] = \sum_{i=0}^{n}[M^{(i,j)}] = \sum_{i=0}^{n}K'^{\,(i,j)}a_1^i\,[M^{(0,j)}] \tag{3.68}$$

Substitution for $[M^{(0,j)}]$ gives

$$[M^{(\bullet,j)}] = \sum_{i=0}^{n}K'^{\,(i,j)}a_1^i \times K^{(0,j)}a_2^j\,[M^{(0,0)}] \tag{3.69}$$

Since

$$K^{(\bullet,j)} = \frac{[M^{(\bullet,j)}]}{[M^{(\bullet,0)}]}\,a_2^j \tag{3.70}$$

one can substitute for $[M^{(0,j)}]$ and $[M^{(\bullet,j)}]$, cancel terms, and finally arrive at

$$K^{(\bullet,j)} = \frac{\sum_{i=0}^{n} K'^{(i,j)} a_1^i}{\sum_{i=0}^{n} K^{(i,0)} a_1^i} \qquad (3.71)$$

Upon forming the derivative of $\ln K^{(\bullet,j)}$ with respect to the logarithm of the activity of L_1, a little algebra gives

$$\frac{\partial \ln K^{(\bullet,j)}}{\partial \ln a_1} = <r_1>_j - <r_1>_0 \qquad (3.72)$$

Here $<r_1>_j$ is the binding density of L_1 when j ligands of L_2 are bound, $<r_1>_0$ is the binding density of L_1 in the absence of any L_2, and the difference between the two is the number of molecules of L_1 exchanged (per macromolecule) upon taking up j molecules of L_2, having started with no L_2 bound at all. Now if experimental conditions can be manipulated such that one can compare the state in which the macromolecule is saturated with respect to species L_2 to the state in which no L_2 is bound at all, then it is possible to deduce the overall stoichiometry of the release (or uptake) of ligand L_1 in response to L_2.

This has found application in explaining, at least partially, the pH sensitivity of hemoglobin's affinity for oxygen. Early work on the Bohr effect in hemoglobin led to the deduction of a shift toward greater acidity (change in pK_a) of a weak acidic group (or groups) on the hemoglobin as it was oxygenated, with concomitant H^+ release (summarized by Edsall [25]). Applying a relation equivalent to Eq. 3.72 above, in connection with their differential titrations of hemoglobin, German and Wyman [26] were able to deduce that at physiological pH about 0.5 protons were released per molecule of O_2 bound (they also found that at acid pH there was a net absorption of protons upon oxygenation). It now appears that in fact several weakly acidic groups are involved in the Bohr effect at physiological pH, principally two histidine residues and the amino terminus of the α polypeptide chain (hemoglobin consists of two α chains and two β chains). When protonated, one of the histidines can make a salt-bridge contact with an aspartate residue, which stabilizes the deoxygenated form of hemoglobin. Hence, acidic pH promotes the loss of oxygen, while alkaline pH will remove the proton, break the salt bridge, and lead to oxygen uptake.

3.4 LINKAGE INVOLVING WEAK AND NONSTOICHIOMETRIC BINDING

There is a very strong tendency to think of binding in terms of site-specific association of a ligand with a macromolecule, a viewpoint that is certainly encouraged by the use of the binding polynomial as a model of the binding system. The binding polynomial approach lends itself well to modeling the formation of well-defined stoichiometric complexes. It is, however, not well-suited to describing cases in which the binding is very weak (such that competition between ligand and solvent for access to the macromolecule dominates the thermodynamics) or the binding is nonstoichiometric (though possibly still strong), as in the association of counterions with highly charged polyions such as DNA. In particular, when studying systems in which the binding is weak, the experimenter must often use relatively high concentrations of ligand (on the order of one molar) in order to obtain an appreciable amount of binding. This might happen, for example, in studying the binding to a protein of urea, of a polyol such as glycerol, or of a salt like sodium sulfate. Now the "binding sites" and their number are not necessarily well defined in terms of particular structural features on the macromolecule; the association here is weak and diffuse. The concentration of the ligand is also high enough, comparable to that of the solvent water, that this species might more properly be regarded as a cosolvent than as a ligand. In these cases it is necessary to return to a macroscopic thermodynamic description of the binding and of the influence of one solution component on the thermodynamic activity of another component. The general linkage concepts, however, can still be applied in these situations, by introducing first the concept of preferential interaction.

3.4.1 PREFERENTIAL INTERACTION

Preferential interaction is perhaps most clearly introduced by considering a dialysis equilibrium experiment, in which a semipermeable membrane separates a solution of ligand plus macromolecule from a solution containing only ligand. This is a three-component system, and this book will use Scatchard's system of subscripts to indicate species or components here [27]: a subscript 1 denotes the solvent (here, water), a subscript 2 denotes the (uncharged) macromolecule species (to avoid some technical thickets it is convenient to set aside ionic solutes and macromolecules and leave their treatment to the specialist), and a subscript 3 denotes an added solute species, which may be regarded provisionally as the ligand species.

Equilibrium dialysis is a standard method for measuring the extent of ligand binding to a biopolymer. Because it is a true equilibrium thermodynamic method, a dialysis experiment will provide a genuinely thermodynamic measurement of the amount of ligand binding, a combination of both site-specific and any nonspecific binding. In principle, given a sufficiently sensitive and accurate method of measuring concentrations, dialysis could measure even quite weak and nonspecific binding, for example, the affinity for proteins of simple inorganic ions like chloride, or of denaturing agents like urea, or of stabilizing agents like sucrose.

Furthermore, the dialysis measurement could (again in principle) provide information about the competition between water and any added solute or cosolvent for binding to the biopolymer, if one were able to follow the accumulation or exclusion of water (or cosolvent or solute) from the chamber containing the biopolymer. However, in the competition between water and these solutes or cosolvents, the situation is one of weak binding (binding constants of 10^3 M^{-1} or less) for all species concerned. Characterization of the binding competition requires accurate and precise measurements of small differences in solute concentration, done at relatively high concentrations of the added solute. Applying equilibrium dialysis to this sort of competition is experimentally difficult [28,29]. Alternative methods include isopiestic measurements [30] and vapor pressure osmometry [31,32]. These latter methods are still quite demanding in terms of experimental technique, and the interpretation of results is technically involved; the details will not be pursued here. For simplicity it seems preferable instead to analyze the classical equilibrium dialysis experiment, which will illustrate the general physical chemical approaches and principles to be used in treating these systems of weak and nonspecific binding.

Consider now a system with three components: solvent, macromolecule (in the inner chamber only, of course), and a third diffusible solute species, and let the system equilibrate. Suppose that one more molecule of component 2 (nondiffusible macromolecule) were added to the inner chamber. This will perturb the equilibrium across the membrane, perhaps drawing some molecules of the solute (component 3) into the inner solution, or perhaps expelling them. There may also be a shift in the number of molecules of solvent across the membrane. Now, if the outer compartment is very large by comparison to the inner one, then it can serve to buffer the chemical potentials of solvent and solute; that is, their chemical potentials will be constant during this experiment.

It becomes convenient to construct a response function for the system in terms of a rate of change of the concentration of component 3 in the inner compartment as component 2 is added. The thought experiment involved the addition of only one molecule of component 2, so that the concentration change is microscopic, infinitesimally small. Thus, it is legitimate to write the response function in terms of an infinitesimal rate of change of m_3 with m_2. This ratio is the *preferential interaction coefficient* for component 2 with component 3:

$$\Gamma_{3,2} = \left(\frac{\partial m_3}{\partial m_2} \right)_{T,\mu_1,\mu_3} = \Gamma_{\mu_1,\mu_3} \tag{3.73}$$

Notice the constraints on the partial derivative: the temperature T (but not the pressure P) is fixed; also, the chemical potentials of solvent and solute are constant, an important factor when converting measurements made under constraints of, say, constant pressure, temperature, and chemical potential of solvent, μ_1. The quantity Γ_{μ_1,μ_3} expresses a thermodynamic restriction on possible simultaneous changes in the amounts of solute and macromolecule in the inner chamber. The molalities m_2 and m_3 cannot be changed arbitrarily, but must change in concert in such a way as to maintain the constancy of the chemical potentials of solvent and solute. The limiting value of Γ_{μ_1,μ_3} for infinite dilution of the macromolecule is $\Gamma^0_{\mu_1,\mu_3}$, which can be evaluated by plotting m_3 as a function of m_2 under the above-noted restrictions on Γ, μ_1 and μ_3. For solutions sufficiently dilute in macromolecule the plot should be linear, with a slope equal to $\Gamma^0_{\mu_1,\mu_3}$.

The quantity Γ_{μ_1,μ_3} expresses the *total* binding of solute by the macromolecule. A *positive* value for Γ_{μ_1,μ_3} indicates a general *accumulation* of a solute species in the vicinity of the macromolecule, by stoichiometric site binding as well as by weak, nonstoichiometric binding. A *negative* value for Γ_{μ_1,μ_3} expresses the *exclusion* of solute from the vicinity of the macromolecule in preference for the solvent (Figure 3.10).

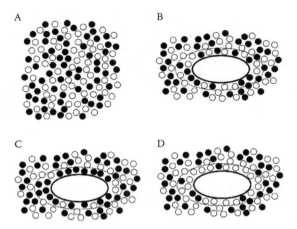

FIGURE 3.10 A schematic representation of the dialysis experiment to define binding of a solute (black dots) to a protein (white oval) in the presence of solvent water (open dots). A. Bulk solvent, no protein present. B. Protein interacting with surrounding shell of solution (corresponding to the inner solution of a dialysis experiment), with no preference for either solvent or solute species (approximately equal numbers of solvent and solute molecules in the shell). C. Protein now preferentially accumulating solute in the shell. D. Protein now preferentially accumulating solvent water in the shell.

The attraction or repulsion of solute or solvent from the region about the macromolecule can derive from a number of sources. Bulky solutes or cosolvents may be excluded simply on the basis of their size; because of the volume they occupy they cannot compete with water for access to small crevices or cavities within the macromolecule, and so they would tend to be excluded. Some solutes may be attracted to polar surfaces on the macromolecule (e.g., urea, as it interacts with exposed amide linkages on a protein), while others may be repelled from such surfaces (steric repulsion, or simple volume occupancy in solution by a solute, can be included here). Local Coulombic attractions or repulsions, involving ionic moieties on the surface, can clearly be important as well. Since typical soluble polymers, especially biopolymers, will present a variety of types of surfaces, it is difficult to predict beforehand what the magnitude of Γ_{μ_1, μ_3} will be.

It is important to note that for $\Gamma_{\mu_1, \mu_3} > 0$ the accumulated solute molecules are not necessarily in physical contact with the surface of the polymer; they are simply more concentrated in the inner chamber than in the outer chamber of the dialysis apparatus. Thus, a dialysis equilibrium, and the preferential interaction coefficient, may indicate more binding than would be detected by, e.g., a spectroscopic assay. Spectroscopic (and other nonthermodynamic) binding assays generally will detect only binding that involves specific sites, since it is usually only close-range contacts that can perturb spectroscopic signals. However, dialysis equilibrium, and determination of the preferential interaction, can detect both the site-specific binding of a ligand by the macromolecule as well as any weak or diffuse binding of solute to the macromolecule.

The extent and kinds of weak binding permitted by this definition may be somewhat surprising. In addition to effects mediated by direct contact of the macromolecule with solute molecules, this definition of binding could include more distant interactions between macromolecule and solute, even the momentary slowing of the translation or rotation of a solute molecule that is not even in contact with the protein surface. Also, from this point of view the volume over which such weak effects extend could be the entire inner chamber of the dialysis apparatus. For example, the dialysis experiment is one route to the detection of an increase in the density of counterions near a polyion, something that would be difficult to determine by, e.g., filtration assays or spectroscopic tests.

There is a simple heuristic connection between the preferential interaction coefficient and the observed accumulation (or expulsion) of water and solute from the inner chamber in the dialysis experiment [33,34]. Imagine that the inner chamber contains a kilogram of solvent, and that the outer chamber is much larger in volume than the inner, so that the composition in this outer chamber does not change. Now add an infinitesimal quantity of macromolecule to the inner chamber (corresponding to an infinitesimal change in molality of macromolecule dm_2) and, upon equilibration, observe that the "free" amounts of solvent and solute in the inner chamber have changed by the amounts $B_1 \, dm_2$ and $B_3 \, dm_2$, respectively. Here B_1 and B_3 are phenomenological parameters (both positive) accounting for the hydration ("solvent binding") and ligation ("solute binding") of the macromolecule. Replacing the kilogram of solvent in the inner chamber with its molality m_1 gives a formula for the preferential interaction coefficient:

$$\left(\frac{\partial m_3}{\partial m_2} \right)_{T,P,\mu_3} = B_3 - \frac{m_3}{m_2} B_1 = \Gamma_{\mu_3} \tag{3.74}$$

Notice that constancy of μ_1 is no longer specified here so that, strictly speaking, this is not the same quantity as Γ_{μ_1,μ_3}. This is, however, a useful form for interpreting macromolecular equilibria. The original derivation of this formula [33,34] makes no reference to binding sites and does not propose any microscopic model of the binding; it is purely a phenomenological derivation based on observed changes in concentrations. Schellman [35] obtained this relation using a site-binding model, and Record and Anderson [36] have also derived the same relation using a two-domain model (essentially a phase-partitioning model) for preferential interactions. Similar relations have been obtained by Reisler et al. [37] and by Tanford [38], so it seems a generally applicable phenomenological formula.

In the original formulation of the preferential interaction coefficient, the chemical potentials of components 1 and 3 were specified as remaining constant. The pressure, however, is not constant in a dialysis experiment because the inner solution experiences a higher pressure than the outer solution. Suppose that it were possible to hold the pressure constant while also fixing the chemical potential of the solvent. These are the experimental conditions for vapor pressure osmometry or isopiestic measurements, which can be used experimentally in characterizing weak binding and preferential interactions. Then a preferential interaction coefficient for these conditions can be defined as

$$\left(\frac{\partial m_3}{\partial m_2}\right)_{T,P,\mu_1} = \Gamma_{\mu_1} \tag{3.75}$$

It is possible to connect Γ_{μ_1} and the related quantity Γ_{μ_1,μ_3} to the quantity Γ_{μ_3}, the form of preferential interaction coefficient that is most useful in interpreting macromolecular equilibria, though some care must be taken in the process (for details see Casassa and Eisenberg [39], Eisenberg [40], Courtenay et al. [32], and Anderson et al. [41,42]). While these different formulations of the preferential interaction may yield very nearly the same numbers when dealing with highly charged polyelectrolytes in concentrated salt solution, they may be quite different for neutral polymers interacting with neutral solutes, as would be the case of proteins interacting with urea, sucrose, or glycerol [41,42]; use caution in choosing formulas to analyze such "neutral" systems.

If hydration interactions with the macromolecule are strong enough, then there can be a net expulsion of the ligand/solute from the inner chamber and a net accumulation of water. This results in what appears to be "negative" binding of the solute. Based on experience with tight-binding systems, Γ_{μ_1,μ_3} might be expected to be positive, or at least nonnegative. In fact, there are numerous instances in which negative values for the preferential interaction, and hence "negative binding," have been found. For example, for the system RNase A with 1 M sucrose, the preferential interaction of the protein with the sugar, as measured by the derivative $(\partial m_3/\partial m_2)$ at constant T, μ_1, and μ_3, is −7.6 [43]; for the same protein in 40% glycerol (v/v), this parameter equals −19.1 [44]. These surprising results seem to imply some sort of repulsion of glycerol or sucrose by the protein, such that in a dialysis experiment the inner solution would contain less of the small solute than it would if no protein were present. Also, the preferential interaction may change sign as the solute concentration rises. For example, Arakawa and coworkers [45] report that in the interaction of $MgCl_2$ with β-lactoglobulin at pH 5.1 (the isoelectric point of the protein, to minimize ionic interactions, either attractive or repulsive), Γ_{μ_1,μ_3} is positive (+0.95) for 0.5 M $MgCl_2$, but negative at 1.0 M ($\Gamma_{\mu_1,\mu_3}=-1.00$), and positive again at 3 M ($\Gamma_{\mu_1,\mu_3}=+3.21$). Notice that quite high concentrations of the ligand are used in all these cases, a reflection of the net weakness of the interactions. The negative value for Γ_{μ_1,μ_3} just describes the net result, that at a particular ligand concentration there is less ligand in the inner dialysis chamber than would be expected if there were simple mixing of ligand across the dialysis membrane. The exclusion of the ligand, described in terms of macroscopic concentrations in the inner dialysis chamber, results in a negative value for $(\partial m_3/\partial m_2)$, or equivalently for Γ_{μ_1,μ_3}. Since water is now the component in the inner chamber that is present in amounts greater than expected for random mixing, these situations are often described as cases of *preferential hydration* of the protein.

It is worth stating again that Γ_{μ_1,μ_3} is a true thermodynamic quantity that includes both stoichiometric and nonstoichiometric binding of solute and water. However, the parameters B_1 and B_3 are not true thermodynamic quantities but heuristic quantities, useful in suggesting points for further investigation and development but not to be treated as fundamental characteristics of the system. For solutes that are specifically and tightly bound, experiments are usually done with dilute solutions of macromolecule and solute, such that the ratio m_3/m_1 is very small; the experimental techniques here (e.g., spectroscopic or calorimetric, etc.) detect only the B_3 contribution to the overall binding. For these tight-binding solutes, the term $(m_3/m_1)B_1$ is typically much smaller than B_3, and Γ_{μ_1, μ_3} will approach B_3, which then would represent the stoichiometric, site-specific binding of the solute. For solutes that are bound only weakly but at a number of sites (as is the case, for example, with urea binding to proteins), $(m_3/m_1)B_1$ can become comparable in magnitude to B_3. In these cases dialysis experiments (measuring the global binding Γ_{μ_1, μ_3}) can easily give different results from spectroscopic or calorimetric titrations (measuring site occupancy, e.g., B_3); the latter measurements would need correcting for hydration effects in order to produce the true extent of binding.

If the solute were completely excluded by the macromolecule ($B_3 = 0$), then presumably the corresponding limiting value of B_1 (denoted as B_1°) would represent the hydration of the macromolecule in the absence of any solute. Given that waters of hydration might be expected to each cover approximately 9 Å^2 on the exposed surface of a protein [46], and that the typical soluble globular protein will have exposed to solvent several thousands of Å^2 of surface, the number B_1° could be rather large. Courtenay et al. [32] have quantitatively characterized the high degree of exclusion of glycine betaine by bovine serum albumin over a range of solute and protein concentrations. Extrapolating from their results with glycine betaine, they estimated a lower bound to B_1° for bovine serum albumin of 2.8×10^3 water molecules per protein molecule. If water were packed (at bulk density) in a monolayer around the surface of the protein, then the corresponding value of B_1° would be approximately 3.2×10^3 water molecules per protein molecule, so their results indicate a strong exclusion of glycine betaine around molecules of this protein.

3.4.2 Preferential Interaction and Macromolecular Equilibria

Previously, the effects of a ligand species L on two sorts of macromolecular equilibria, $A \rightleftharpoons B$ and $A + B \rightleftharpoons C$, were described in terms of a set of binding polynomials, with the result that $\partial \ln K_{obs}/\partial \ln a_L = \Delta < r_L >$. When there are changes in the activity of the solvent water and/or in the activity of other solutes present, etc., then the more general relation $\partial \ln K_{obs}/\partial \ln a_L = \Delta < r_L > + (\Delta < r_W > \times \partial \ln a_W/\partial \ln a_L)$ should be used instead. Both of these relations assume that a degree of saturation Y can be defined for each of the macromolecular species, and that there is a well-defined stoichiometric uptake or release of the ligand; that is, the term $\Delta <r_L>$ should be an integer that could be recovered from a suitable log-log plot. As discussed before, the second term on the right-hand side in the latter relation accounts for hydration effects as well as the indirect effect of the ligand species on the activity of the solvent water. Any appreciable contribution from this term may result in a plot of $\ln K_{app}$ versus $\ln a_L$, indicating apparent nonstoichiometric binding, particularly in concentrated solutions.

If the solute species 3 does not bind to specific sites, but is instead more loosely bound (perhaps in a delocalized way, as with small inorganic cations associated with a large polyion like DNA), can a log-log plot say something about the binding stoichiometry? This will require a reformulation of the relation for $\partial \ln K_{obs}/\partial \ln a_L$ in terms of preferential interactions and not as functions of binding polynomials and their implied site-binding stoichiometry.

The route to the basic formula is simple and a nice example of thermodynamic reasoning. The present discussion, however, will be limited to the binding of a general neutral solute species 3 with the neutral macromolecular species. The interactions of ions with polyelectrolytes, though of great practical and theoretical interest, requires special handling of thermodynamic activities [47], and only a summary of results from that treatment will be given here. The route starts with a definition of the observed equilibrium constant for a macromolecule (component 2) equilibrating between conformations A and B, using activities here and not simple concentrations:

$$\ln K_{obs} = \ln a_2^B - \ln a_2^A \tag{3.76}$$

so

$$\left(\frac{\partial \ln K_{obs}}{\partial \ln a_3} \right)_{T,P,m_2} = \left(\frac{\partial \ln a_2^B}{\partial \ln a_3} \right)_{T,P,m_2} - \left(\frac{\partial \ln a_2^A}{\partial \ln a_3} \right)_{T,P,m_2} \tag{3.77}$$

(the superscripts of A and B on the activity a_2 are to indicate the conformation of component 2). For either form of the macromolecule, A or B, there is the relation between chemical potential and activity:

$$\left(\frac{\partial \ln a_2}{\partial \ln a_3} \right)_{T,P,m_2} = \left(\frac{\partial \mu_2}{\partial \mu_3} \right)_{T,P,m_2} = \left(\frac{\partial \mu_2/\partial m_3}{\partial \mu_3/\partial m_3} \right)_{T,P,m_2} \tag{3.78}$$

The exactness of these partial derivatives gives

$$\left(\frac{\partial \mu_3}{\partial m_2} \right)_{T,P,m_3} = \left(\frac{\partial \mu_2}{\partial m_3} \right)_{T,P,m_2} \tag{3.79}$$

so

$$\left(\frac{\partial \ln a_2}{\partial \ln a_3} \right)_{T,P,m_2} = \frac{(\partial \mu_3/\partial m_2)_{T,P,m_3}}{(\partial \mu_3/\partial m_3)_{T,P,m_2}} \tag{3.80}$$

Application of the cyclic permutation rule gives

$$\left(\frac{\partial m_3}{\partial m_2} \right)_{T,P,\mu_3} = - \frac{(\partial \mu_3/\partial m_2)_{T,P,m_3}}{(\partial \mu_3/\partial m_3)_{T,P,m_2}} \tag{3.81}$$

Then, substituting this into Equation 3.80, for either A or B there is

$$\left(\frac{\partial \ln a_2}{\partial \ln a_3} \right)_{T,P,m_2} = \left(\frac{\partial \mu_2}{\partial \mu_3} \right)_{T,P,m_2} = - \left(\frac{\partial m_3}{\partial m_2} \right)_{T,P,\mu_3} \tag{3.82}$$

so that finally

$$\left(\frac{\partial \ln K_{obs}}{\partial \ln a_3}\right)_{T,p,\mu_3} = \Gamma^B_{\mu_3} - \Gamma^A_{\mu_3} = \Delta\Gamma_{\mu_3} \qquad (3.83)$$

The extension to treat the equilibrium $A + B \rightleftharpoons C$ is straightforward, with only a slight generalization of the right side of Eq. 3.83 to $\Delta\Gamma_{\mu_3} = \Gamma^C_{\mu_3} - \Gamma^A_{\mu_3} - \Gamma^B_{\mu_3}$.

Notice that the preferential interaction coefficient here is that for constancy of the chemical potential of component 3, at constant pressure. This is not the same as the coefficient measured in a dialysis equilibrium, which is Γ_{μ_1,μ_3}. Γ_{μ_3} is the form most readily applicable to these macromolecular equilibria, but unfortunately it is not directly accessible by experiment. Also, the thermodynamics become more complicated when a second small molecule solute is added to the mixture, as when a simple salt is mixed with a urea or glycerol solution; a fairly extensive set of measurements is then needed. The requirements have been explored in a series of recent papers from the Record group [31,41,42,48,49].

Record et al. [47] give a more extended and rigorous treatment of preferential interactions that considers electrolytes and polyelectrolytes, and they specifically treat the important case of the binding of a charged ligand to a polyelectrolyte in the presence of a simple 1:1 electrolyte. The equilibrium here may be written as

$$L^{Z_L} + DNA \text{ site (D)} \rightleftharpoons \text{complex (LD)} \qquad (3.84)$$

To summarize their results, the variation of K_{obs} with simple salt can be written as

$$\frac{\partial \ln K_{obs}}{\partial \ln a_\pm} = (|Z_{LD}| + 2\Gamma_{3,LD}) - (|Z_D| + 2\Gamma_{3,D}) - (|Z_L| + 2\Gamma_{3,L}) \qquad (3.85)$$

The quantities Z_i are ionic valences and $|Z_i|$ is the absolute value of the valence.

Each of the terms in parentheses represents the amount of binding (the "thermodynamic extent of association" as Record et al. put it) of the small ions with the indicated species; the terms include both site-bound and nonspecifically bound ions. For DNA at low concentrations, in dilute salt solutions, the preferential interaction of DNA with salt is dominated by the cation-DNA interaction. Thus, the quantity $(|Z_{LD}| + 2\Gamma_{3,LD}) - (|Z_D| + 2\Gamma_{3,D})$ can be interpreted as the thermodynamic extent of release of cations (both site-bound and nonspecifically bound) from the DNA, and $(|Z_L| + 2\Gamma_{3,L})$ is the amount of cation release attributable to the ligand species L.

As an approximation, the region on the DNA occupied by the ligand L can be considered to differ from the corresponding region on the unligated DNA by the elimination of $|Z_L|$ DNA phosphate charges in the bound region. One might think of this as each individual charge on the ligand neutralizing a single charged phosphate group as the ligand binds to the DNA, with the consequent release of thermodynamically bound counterions into bulk solution. The released counterions gain a great deal of entropic freedom upon being released, and it is this entropy change that is the primary driving force behind the attachment of the ligand to the polyion.

At low salt concentrations, in which so-called limiting laws describe the thermodynamic activity of the ions, the preferential interaction terms collapse, leaving

$$\frac{\partial \ln K_{obs}}{\partial \ln a_{\pm}} = - |Z_L| \left(1 - \frac{1}{2\xi_D} \right) \tag{3.86}$$

The quantity ξ_D is the charge density parameter for the polyion, defined by

$$\xi_D = \frac{e^2}{\varepsilon k_B T b} \tag{3.87}$$

where e is the charge on the proton, ε is the dielectric constant, T is the temperature,
and b is the axial charge spacing on the polyion. For double-stranded DNA, ξ has the
numerical value of 4.2, while for single-stranded DNA it is about 2.1.

Equation 3.86 predicts that the plot of $\ln K_{obs}$ as a function of the logarithm of
salt activity should be approximately linear, with a (negative) slope proportional to
the charge on the ligand (the part that neutralizes polyion charges within the binding site), and a dependence on the density of charge on the polyion. A wide variety
of DNA-binding systems show just this sort of behavior, over a surprisingly wide
range of salt concentrations. An example is shown below in Figure 3.11, taken from

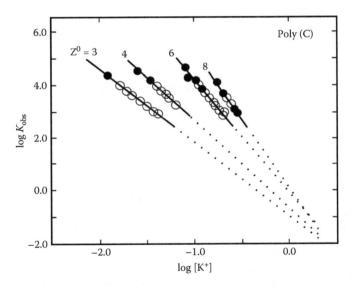

FIGURE 3.11 Binding of oligolysines to single-stranded poly(rC). The open and filled circles
indicate results obtained by slightly different titration techniques. Adapted with permission
from *Biochemistry* Vol. 32, D.P. Mascotti and T.M. Lohman, "Thermodynamics of single-
stranded RNA and DNA interactions with oligolysines containing tryptophan. Effects of base
composition," pages 10568–10579. Copyright 1993 American Chemical Society.

the study of Mascotti and Lohman [50] on the binding of a series of oligolysines to single-stranded polynucleotides.

Here the ligands, which contain a single tryptophan residue as a fluorescent tag and which carry charges of +3, +4, +6, and +8, respectively, are binding to the single-stranded polynucleotide, poly(rC). As expected from theory, the observed binding constant does indeed decline steeply with increasing salt, and the various log-log plots are linear (at least over the salt concentration range explored). Also, the slopes of the plots track at least approximately with the magnitude of the charge on the oligolysine ligand. Furthermore, at one molar salt concentration, the curves extrapolate to a value near unity for K_{obs}, as expected for a binding driven primarily by the release of counterions from the DNA.

REFERENCES

1. Wyman, J. and Gill, S.J., *Binding and Linkage*. Mill Valley, CA: University Science Books, 1990.
2. Wyman, J., Linked functions and reciprocal effects in hemoglobin: A second look, *Adv. Prot. Chem.* 19, 223, 1964.
3. Wyman, J., Heme proteins, *Adv. Prot. Chem.* 4, 407, 1948.
4. Goldstein, A. and Barrett, R.W., Ligand dissociation constants from competition binding assays: Errors associated with ligand depletion, *Molec. Pharmacol.* 31, 603, 1987.
5. Cheng, C. and Prusoff, W.H., Relationship between the inhibition constant (K_I) and the concentration of inhibitor which causes 50 per cent inhibition (I_{50}) of an enzymatic reaction, *Biochem. Pharmacol.* 22, 3099, 1973.
6. Contreras, M.L., Wolfe, B.B., and Molinoff, P.B., Thermodynamic properties of agonist interactions with the *beta* adrenergic receptor-coupled adenylate cyclase system. I. High- and low-affinity states of agonist binding to membrane-bound *beta* adrenergic receptors, *J. Pharmacol. Exper. Therap.* 237, 154, 1986.
7. Horovitz, A. and Levitzki, A., An accurate method for determination of receptor-ligand and enzyme-inhibitor dissociation constants from displacement curves, *Proc. Natl. Acad. Sci. USA* 84, 6654, 1987.
8. Almagor, H. and Levitzki, A., Analytical determination of receptor-ligand dissociation constants of two populations of receptors from displacement curves, *Proc. Natl. Acad. Sci. USA* 87, 6482, 1990.
9. Henis, Y.I. and Levitzki, A., Ligand competition curves as a diagnostic tool for delineating the nature of site-site interactions: Theory, *Eur. J. Biochem.* 102, 449, 1979.
10. Jezewska, M.J. and Bujalowski, W., A general method of analysis of ligand binding to competing macromolecules using the spectroscopic signal originating from a reference macromolecule. Application to *Escherichia coli* replicative helicase DnaB protein-nucleic acid interactions, *Biochemistry* 35, 2117, 1996.
11. Zot, A.S. and Potter, J.D., Structural aspects of troponin-tropomyosin regulation of skeletal muscle contraction, *Annu. Rev. Biophys. Biophys. Chem.* 16, 535, 1987.
12. Kolb, A. et al., Transcriptional regulation by cAMP and its receptor protein, *Annu. Rev. Biochem.* 62, 749, 1993.
13. Bolin, J.T. et al., Crystal structures of *Escherichia coli* and *Lactobacillus casei* dihydrofolate reductase refined at 1.7 Å resolution, *J. Biol. Chem.* 257, 13650, 1982.
14. Blakley, R.L. and Cocco, L., Binding of methotrexate to dihydrofolate reductase and its relation to protonation of the ligand, *Biochemistry* 24, 4704, 1985.

15. Kuyper, L.F. et al., Receptor-based design of dihydrofolate reductase inhibitors: Comparison of crystallographically determined enzyme binding with enzyme affinity in a series of carboxy-substituted trimethoprim analogues, *J. Med. Chem.* 28, 303, 1985.

16. Anderson, D.G., Hammes, G.G., and Walz, F.G., Binding of phosphate ligands to ribonuclease A, *Biochemistry* 7, 1637, 1968.

17. Flogel, M. and Biltonen, R.L., The pH dependence of the thermodynamics of the interaction of 3′-cytidine monophosphate with ribonuclease A, *Biochemistry* 14, 2610, 1975.

18. Record, M.T. Jr., Anderson, C.F., and Lohman, T.L., Thermodynamic analysis of ion effects on the binding and conformational equilibria of proteins and nucleic acids: The roles of ion association or release, screening, and ion effects on water activity, *Q. Rev. Biophys.* 11, 103, 1978.

19. Aune, K.C., Goldsmith, L.C., and Timasheff, S.N., Dimerization of α-chymotrypsin. II. Ionic strength and temperature dependence, *Biochemistry* 10, 1617, 1971.

20. Aune, K.C. and Timasheff, S.N., Dimerization of α-chymotrypsin. I. pH dependence in the acid region, *Biochemistry* 10, 1609, 1971.

21. Friedberg, F. and Bose, S., Ion binding by α-chymotrypsin, *Biochemistry* 8, 2564, 1969.

22. Haire, R.N. and Hedlund, B.E., Thermodynamic aspects of the linkage between binding of chloride and oxygen to human hemoglobin, *Proc. Natl. Acad. Sci. USA* 74, 4135, 1977.

23. Colombo, M.F., Rau, D.C., and Parsegian, V.A., Reevaluation of chloride's regulation of hemoglobin oxygen uptake: The neglected contribution of protein hydration in allosterism, *Proc. Natl. Acad. Sci. USA* 91, 10517, 1994.

24. Colombo, M.F., Rau, D.C., and Parsegian, V.A., Protein solvation in allosteric regulation: A water effect on hemoglobin, *Science* 256, 655, 1992.

25. Edsall, J.T., Hemoglobin and the origins of the concept of allosterism, *Fed. Proc.* 39, 226, 1980.

26. German, B. and Wyman, J., The titration curves of oxygenated and reduced hemoglobin, *J. Biol. Chem.* 117, 533, 1937.

27. Scatchard, G., Physical chemistry of protein solutions. I. Derivation of the equations for the osmotic pressure, *J. Am. Chem. Soc.* 68, 2315, 1946.

28. Scatchard, G., Batchelder, A.C., and Brown, A., Preparation and properties of serum and plasma proteins. VI. Osmotic equilibria in solutions of serum albumin and sodium chloride, *J. Am. Chem. Soc.* 68, 2320, 1946.

29. Klotz, I.M., Ligand-protein binding affinities, in *Protein Function: A Practical Approach*, Creighton, T.E., Ed. Oxford: IRL Press, 1989, p. 25.

30. Hade, E.P.K. and Tanford, C., Isopiestic compositions as a measure of preferential interactions of macromolecules in two-component solvents. Application to proteins in concentrated aqueous cesium chloride and guanidine hydrochloride, *J. Am. Chem. Soc.* 89, 5034, 1967.

31. Zhang, W. et al., Thermodynamic characterization of interactions of native bovine serum albumin with highly excluded (glycine betaine) and moderately accumulated (urea) solutes by a novel application of vapor pressure osmometry, *Biochemistry* 35, 10506, 1996.

32. Courtenay, E.S. et al., Vapor pressure osmometry studies of osmolyte-protein interactions: Implications for the action of osmoprotectants in vivo and for the interpretation of 'osmotic stress' experiments in vitro, *Biochemistry* 39, 4455, 2000.

33. Timasheff, S.N. and Inoue, H., Preferential binding of solvent components to proteins in mixed water-organic solvent system, *Biochemistry* 7, 2501, 1968.

34. Inoue, H. and Timasheff, S.N., Preferential and absolute interactions of solvent components with proteins in mixed solvent systems, *Biopolymers* 11, 737, 1972.

35. Schellman, J.A., A simple model for solvation in mixed solvents. Applications to the stabilization and destabilization of macromolecular structures, *Biophys. Chem.* 37, 121, 1990.

36. Record. M.T. Jr. and Anderson, C.F., Interpretation of preferential interaction coefficients of nonelectrolytes and of electrolyte ions in terms of a two domain model, *Biophys. J.* 68, 786, 1995.

37. Reisler, E., Haik, Y., and Eisenberg, H., Bovine serum albumin in aqueous guanidine hydrochloride solutions. Preferential and absolute interactions and comparison with other systems, *Biochemistry* 16, 197, 1977.

38. Tanford, C., Extension of the theory of linked functions to incorporate the effects of protein hydration, *J. Mol. Biol.* 39, 539, 1969.

39. Casassa, E.F. and Eisenberg, H., Thermodynamic analysis of multicomponent solutions, *Adv. Prot. Chem.* 19, 287, 1964.

40. Eisenberg, H., *Biological Macromolecules and Polyelectrolytes in Solution.* Oxford: Clarendon Press, 1976.

41. Anderson, C.F., Courtenay, E.S., and Record, M.T. Jr., Thermodynamic expressions relating different types of preferential interaction coefficients in solutions containing two solute components, *J. Phys. Chem. B* 106, 418, 2001.

42. Anderson, C.F. et al., Generalized derivation of an exact relationship linking different coefficients that characterize thermodynamic effects of preferential interactions, *Biophys. Chem.* 101–102, 497, 2002.

43. Lee, J.C. and Timasheff, S.N., The stabilization of proteins by sucrose, *J. Biol. Chem.* 256, 7193, 1981.

44. Gekko, K. and Timasheff, S.N., Mechanism of protein stabilization by glycerol: Preferential hydration in glycerol-water mixtures, *Biochemistry* 20, 4667, 1981.

45. Arakawa, T., Bhat, R., and Timasheff, S.N., Preferential interactions determine protein solubility in three-component solutions: The $MgCl_2$ system, *Biochemistry* 29, 1914, 1990.

46. Gill, S.J. et al., Anomalous heat capacity of hydrophobic solvation, *J. Phys. Chem.* 89, 3758, 1985.

47. Record, M.T. Jr., Zhang, W., and Anderson, C.F., Analysis of effects of salts and uncharged solutes on protein and nucleic acid equilibria and processes: A practical guide to recognizing and interpreting polyelectrolyte effects, Hofmeister effects and osmotic effects of salts, *Adv. Protein Chem.* 51, 281, 1998.

48. Courtenay, E.S., Capp, M.W., and Record, M.T. Jr., Thermodynamics of interactions of urea and guanidinium salts with protein surface: Relationship between solute effects on protein processes and changes in water-accessible surface area, *Protein Sci.* 10, 2485, 2001.

49. Hong, J. et al., Preferential interactions in aqueous solutions of urea and KCl, *Biophys. Chem.* 105, 517, 2003.

50. Mascotti, D.P. and Lohman, T.M., Thermodynamics of single-stranded RNA and DNA interactions with oligolysines containing tryptophan. Effects of base composition, *Biochemistry* 32, 10568, 1993.

4 Cooperativity

This chapter will give the basics of how to describe and model cooperative interactions in macromolecular binding. Cooperativity calls for at least two binding sites that can interact with one another, so the discussion here will be concerned with multisite systems exhibiting interdependency in their binding action. This will include some basic structural models for cooperative ligand binding to single- and multisubunit receptors. Chapter 5 will take up cooperative binding to linear polymers and planar arrays of binding sites.

4.1 THE PHENOMENON OF BINDING COOPERATIVITY

In a plot of $<r>$ as a function of [L], the simple mass action law for bimolecular complex formation leads to a rectangular hyperbola for the binding curve. Binding curves in many biological systems, however, do not follow this simple behavior. Often the binding curve appears to be sharper, with the binding density rising to saturation more quickly than would be expected by the law of mass action. Perhaps there is also a sort of initial "lag phase" to the curve, in which adding ligand seems to have little or no effect on the binding density. The result is a sigmoid curve where a hyperbola might be expected. Also, when the data are plotted according to standard formulations that should yield straight lines if the sites were equal and independent (e.g., Scatchard, double-reciprocal, etc.), the data instead trace a line that curves; for example, the data in a Scatchard plot may form a "hump," or a Hill plot may exhibit a slanted S-shaped trend in the data. This may then be a case of *binding cooperativity*.

4.1.1 COOPERATIVE BINDING IN THE OXYGENATION OF HEMOGLOBIN

Probably the best-known, and certainly one of the most closely studied, examples of biochemical binding cooperativity is the oxygenation of hemoglobin. Observations of cooperativity in this system have quite a long history. In 1904 Bohr and colleagues reported oxygenation curves for dog blood that had a definite sigmoid character: oxygen binding at first rises slowly with increasing oxygen partial pressure, then rises more steeply at intermediate pressures, and finally approaches a plateau of saturation [1]. This behavior is clearly not that expected from a collection of equal and independent oxygen-binding sites in the blood.

Working with purified protein, T. Svedberg and R. Fåhreus [2], and independently G. S. Adair [3], established hemoglobin's molecular weight and the presence of four oxygen binding sites per molecule. Adair [4] proposed that the oxygenation curves (Figure 4.1) could be described with a ratio of fourth degree polynomials in the partial pressure of oxygen (the partial pressure is proportional to the concentration of dissolved molecular oxygen). In his paper Adair did not explicitly derive his oxygenation

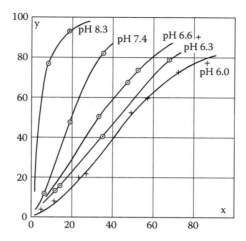

FIGURE 4.1 Cooperativity in the oxygenation of hemoglobin was apparent in early studies. Note the characteristic sigmoid shape of the curves of the percentage of oxygenation as a function of oxygen partial pressure (mm Hg). Adapted with permission from *Journal of Biological Chemistry* Vol. 63, G.S. Adair, "The hemoglobin system. VI. The oxygen dissociation curve of hemoglobin," pages 529–545. Copyright 1925 American Society for Biochemistry and Molecular Biology.

equation, but it is easy to see that he obtained it from a basic binding polynomial. Using current notation the basic Adair binding polynomial can be written as

$$\mathbf{P} = 1 + K^{(1)}[O_2] + K^{(2)}[O_2]^2 + K^{(3)}[O_2]^3 + K^{(4)}[O_2]^4 \tag{4.1}$$

where dissolved O_2 concentrations have replaced oxygen partial pressures. The fourth degree of the polynomial reflects the presence of four binding sites for oxygen in hemoglobin. The constants $K^{(1)}$, ... , $K^{(4)}$ are phenomenological binding constants for forming the species with 1, 2, 3, or 4 oxygen molecules bound, directly from the unligated (completely deoxygenated) hemoglobin. The S-shape of the binding curve results from $K^{(4)}$ having a value substantially greater than that for $K^{(1)}$, when corrected for statistical factors (see Section 4.3.1).

The sigmoid character of the oxygen binding curve indicates some sort of interaction between the binding sites that make it increasingly favorable to bind oxygen. To explain this, Pauling [5] proposed a model for hemoglobin that assumed four equivalent hemes, interacting as pairs and at short range. He considered two arrangements of the heme groups, the first with them at the corners of a square with pairwise interactions along the sides of the square, and the second with the hemes arranged at the apices of a tetrahedron, with interactions now along the vertices of the tetrahedron. Judged by eye, either model gave a good fit to the data (no statistical curve fitting was done). The agreement of the models with the data indicated that the pairwise interactions were favorable and that each contributed about 4 kJ/mol of free energy toward stabilization of the complex. Pauling argued that with this strong an interaction, the heme groups could not lie far from one another, and so he assumed that the hemes must interact directly with one another. Then, since the tetrahedral model required a greater separation of hemes than the square planar model, he favored a square planar arrangement of the protein units so that the interacting units would be closer together. Unfortunately, Pauling was wrong about both the arrangement of the hemes and how they interacted, as the crystal structures of Max Perutz and coworkers later showed quite clearly [6–8]. The heme-containing subunits pack as a tetrahedral unit, not as a planar square, and there is a heme-heme separation of about 25 Å. This is too great a distance for the hemes to interact directly with one another. Instead, the interaction between these relatively distant sites is propagated by rather small shifts in the packing of amino acids within and at the interfaces of the protein subunits. Pauling's reasoning was sound enough, but at the time he proposed his model there was no precedent for the subtlety of the site-site interactions later found in hemoglobin.

4.1.2 COOPERATIVITY IN ENZYME ACTION

During the 1960s, unusual binding isotherms or patterns of catalytic activation were found for a number of complex and metabolically important enzymes. These enzymes did not follow the standard Michaelis-Menten model for enzyme activity, but instead showed S-shaped activation or binding curves, resembling the oxygenation curve for hemoglobin. As with hemoglobin, these sigmoid curves implied there were strong interactions between the sites. A particularly striking example was found with the enzyme aspartate transcarbamoylase. This enzyme showed a markedly sigmoid activation curve as increasing amounts of substrate (aspartate) were added [9]. Furthermore, changes in activity could be produced by adding small amounts of different nucleotide triphosphates (Figure 4.2). ATP could activate the enzyme, shifting the

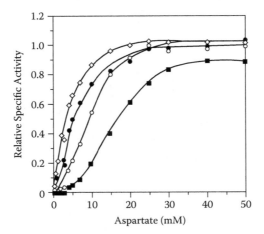

FIGURE 4.2 Cooperativity and allosteric effector action in the carbamoylation of aspartate by aspartate transcarbamoylase. The sigmoid dependence of reaction velocity on substrate concentration indicates cooperative binding of substrate, in the absence of any allosteric effector species (●). Addition of ATP at 2 mM concentration increases affinity for substrate, while addition of either CTP (○) or CTP + UTP (■) at 2 mM concentrations induces a shift to lower affinity for substrate. Adapted from *Proceedings of the National Academy of Sciences USA,* Vol. 86, J.R. Wild et al., "In the presence of CTP, UTP becomes an allosteric inhibitor of aspartate transcarbamoylase," pages 46–50. Copyright 1989 National Academy of Sciences U.S.A.

velocity-versus-substrate concentration curve to the left toward lower substrate concentrations and reducing the degree of sigmoidicity. CTP, on the other hand, shifted the activity curve to the right and it augmented the sigmoid character of the plot.

Additionally, the enzyme underwent gross conformational changes upon binding certain ligands, as shown by X-ray diffraction studies. Structural studies have nicely correlated activity changes with shifts in conformation [10,11]. Figure 4.3 schematically illustrates the symmetric multisubunit nature of the enzyme and the drastic cooperative conformational change between active and inactive forms.

In 1965, Jacques Monod, Jeffries Wyman, and Pierre Changeux proposed a "plausible model" for enzymes with multiple sites or subunits, to explain and correlate the various observations on enzyme activity as a function of substrate concentration, the effects of different inhibitors and activators, and changes in protein conformation [12]. This is the famous *allosteric* model of cooperativity. A distinct but related model involving a *sequential* change in macromolecule conformation was proposed by Daniel Koshland and coworkers in 1966 [13]. Since then these two models have deeply influenced thinking about cooperativity, multiple subunits, and

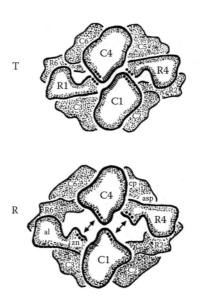

FIGURE 4.3 Aspartate transcarbamoylase structure in a highly schematized illustration. In the absence of the activator ATP, the assembly of six catalytic subunits (denoted as *c* here) form two layers of three subunits each. These layers are linked by three dimeric regulatory subunits (denoted by *r*) to form a compact structure (T state) with lower catalytic activity. Upon binding ATP, the dimeric regulatory subunits rotate and move the two trimeric catalytic layers apart, leading to greater catalytic activity (R state). Adapted with permission from *Biochemistry* Vol. 29, J.E. Gouax and W.N. Lipscomb, "Crystal structures of phosphonoacetamide ligated T and phosphonacetamide and malonate ligated R states of aspartate carbamoyltransferase at 2.8-Å resolution and neutral pH," pages 389–402. Copyright 1990 American Chemical Society.

binding of small molecule inhibitors and activators, not only for enzymes, but for drug-receptor interactions, control of DNA-protein interactions, and other activity.

As a technical aside, it is possible that with certain enzymes, positive cooperativity can arise kinetically. Many enzymatic reactions have both slow and fast steps, or require binding of multiple substrates to occur in a particular order, and these systems may show kinetic cooperativity. Assays for enzyme activity with systems that have multiple substrates are often done under special conditions so as to follow, for example, the steady-state turnover of substrate to product. The phenomenological rate laws for such systems may have one or more terms in the concentration of substrate whose powers or exponents exceed the number of subunits of the enzyme. The phenomenological result is an augmentation of the rate beyond that for an enzyme with independent, noninteracting sites; that is, the system shows positive (kinetic) cooperativity. This book is concerned, however, with binding equilibrium, and the reader interested in further kinetic discussions is referred to Cornish-Bowden [14].

4.2 TERMINOLOGY FOR COOPERATIVE INTERACTIONS

There are two qualitatively distinguishable kinds of cooperativity. *Positive cooperativity* occurs when the binding of one ligand promotes the binding of yet more ligand, beyond the level of uptake expected for independent action by the binding sites. The binding isotherm for a positively cooperative system is steeper than that for a system with independent sites. The qualification regarding independence of the sites is important, as will appear in Section 4.3.

The converse of positive cooperativity is of course *negative cooperativity* (sometimes also called *anticooperativity*) which occurs when binding of one ligand hinders further uptake, again by comparison to independently acting sites. With negative cooperativity the binding curve rises toward saturation more slowly than for a system of independent sites. Negative cooperativity manifests itself through a progressive loss of binding affinity within the same class of sites, which is something quite apart from having multiple independent classes of sites with constant affinity for the ligand, and that may also differ in number of binding sites. (An aside on terminology seems appropriate here: a system with *negative cooperativity* is *not* the same as one that is *noncooperative*. A noncooperative system is one that simply is not cooperative in either the positive or negative sense; the sites are acting independently.)

Many of the enzymes that control the rate of key metabolic processes are sensitive to the concentration of small-molecule metabolites related to the pathway concerned. Presumably, these small molecules are modulating pathway activity by binding to the enzymes in the pathway. In some cases the small molecule modulators resemble a natural substrate of the enzyme concerned; in other cases there is little or no resemblance to substrate. Monod and coworkers coined a set of terms to distinguish between these two different classes of modulators, based on their resemblance to substrate: *isosteric effectors* refers to those modulators of enzyme activity that are structurally akin to substrate, while *allosteric effectors* refers to the class of modulators that do not look much like substrate at all [15,16]. Because of their lack of resemblance to substrate, allosteric effectors were thought to bind at sites *other* than the active site, and that once bound, they produced their effects by altering the enzyme's conformation at the active site. An *allosteric activator* would raise the activity of the enzyme, while an *allosteric inhibitor* would of course lower it. In terms of observable enzyme kinetics, it would be difficult to distinguish an isosteric inhibitor from an allosteric inhibitor; both types of molecules would produce competitive inhibition of the enzyme. The term *isosteric effector* is not frequently used today, but the term *allosteric effector* and its cousin *allosterism* have proven quite durable. In current usage, *allosterism* refers to regulatory effects stemming from conformational changes in the macromolecule as it binds or releases ligands.

Many regulatory enzymes are made up of multiple subunits and have multiple catalytic sites, and they very likely will have multiple sites for binding allosteric inhibitors or activators as well. Often these enzymes are found to undergo changes in conformation or state of aggregation. Because there are multiple binding sites, the possibility for cooperative interactions among the binding and catalytic sites arises; cooperativity of course requires the interaction of two or more sites. The binding of one substrate molecule could favorably influence the binding of more substrate, an example of positive cooperativity. Negative cooperativity induced by binding substrate is also possible. Furthermore, the uptake of allosteric effector molecules could be either positively cooperative or negatively so. There might also be a positive or negative cooperative influence of the effector on substrate uptake (and vice versa). To distinguish among these different results, Wyman [17], and later Monod et al. [12], proposed the term *homotropic effects* to describe the binding interaction among identical ligands, and *heterotropic effects* for the interaction of dissimilar ligands. Also, the second species would be called an *effector* of the binding of the first species, and vice versa.

The sigmoid binding curve (in the direct plot format) that is characteristic of positive cooperativity cannot be described by a binding polynomial that is linear in the ligand concentration. The curvature instead indicates that the binding polynomial must contain terms proportional to the second or higher power of the ligand concentration. Similarly, where there is progressive inhibition of binding as the binding density increases, the binding polynomial again must contain higher order terms. Binding takes place in these systems at two or more sites on the same receptor; there *must* be more than one site in order to produce the higher-than-first-power dependence of the isotherm on the ligand concentration. Homotropic cooperativity depends on the receptor having multiple interacting sites that bind the same species, and it is characterized by a proportionality of the binding to a second or higher power of the concentration of that particular species. This is not necessarily true of heterotropic cooperativity, however. For example, there might be a receptor with two different binding sites, one for substrate and the other for an allosteric effector. The binding polynomial would then be linear in the concentrations of either species.

4.3 CRITERIA FOR COOPERATIVITY IN LIGAND BINDING

In order to define cooperativity in a quantitative way, it is important to first consider what it means to have *no* cooperativity in a multisite system. There are, in fact, some interesting statistical effects here at the microscopic level that might incautiously be interpreted as *negative* cooperativity. Further insight can be gained by looking at a system with several equivalent sites but in which the binding is all or none, an extreme example of positive cooperativity. This leads to an operational definition of cooperativity in terms of the concentration dependence of an effective equilibrium constant for ligand binding. This definition can be developed further by using the binding polynomial's mathematical properties to describe linkage relations in the distribution of the number and type of the bound ligands.

4.3.1 STATISTICAL EFFECTS IN MULTISITE BINDING

The Langmuir binding isotherm model is the prototype for noncooperative binding. As discussed in Chapter 2, it represents the behavior of a system of *equal* and *independent* binding sites, each with binding constant k. Binding of the ligand species occurs at random, with equal a priori probability at any vacant site. The isotherm equation for a system with n sites is compactly expressed as

$$<r> = \frac{n\,k\,[L]}{1+k\,[L]} \tag{4.2}$$

but it can also be written in terms of a collection of phenomenological binding constants:

$$<r> = \frac{K^{(1)}[L] + 2K^{(2)}[L]^2 + 3K^{(3)}[L]^3 + \cdots + nK^{(n)}[L]^n}{1 + K^{(1)}[L] + K^{(2)}[L]^2 + K^{(3)}[L]^3 + \cdots + K^{(n)}[L]^n} \tag{4.3}$$

Another form of the isotherm equation that will be useful in analyzing Hill plots is

$$\frac{Y}{1-Y} = \frac{\sum_{i=1}^{n} iK^{(i)}[L]^i}{n\sum_{i=1}^{n} K^{(i)}[L]^i - \sum_{i=1}^{n} iK^{(i)}[L]^i} \tag{4.4}$$

The connections between the site binding constant k and the phenomenological binding constants are

$$K^{(1)} = \frac{n!}{1!(n-1)!}k = n\,k \tag{4.5}$$

$$K^{(2)} = \frac{n!}{2!(n-2)!}k^2 = \frac{n(n-1)}{2}k^2 \tag{4.6}$$

$$\cdots$$

$$K^{(n)} = \frac{n!}{n!(n-n)!}k^n = k^n \tag{4.7}$$

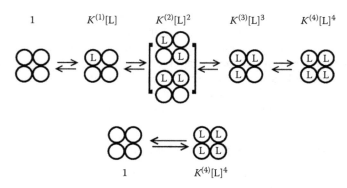

FIGURE 4.4 Comparison of the equal-and-independent-sites model to the all-or-none model for a system with four sites. The independent-sites model includes a number of intermediate states of ligation of the macromolecule, while the all-or-none model has only two states, completely vacant or completely saturated.

The phenomenological binding constants are all expressible in terms of a single underlying site binding constant k, raised to an appropriate power and multiplied by a numerical factor that corrects for the statistically random distribution of the ligands over the available sites. The resulting binding curve is a rectangular hyperbola, without any sigmoid character. Figure 4.4 illustrates the binding states and corresponding statistical weights for a four-site model, comparing the model of progressive one-at-a-time uptake of ligand to the all-or-none uptake model (see Section 4.3.2).

The formulas show that there is a steady decrease in the ratio of $K^{(i)}$ to $K^{(i-1)}$ for systems with higher and higher numbers of occupied sites. For a system with four independent sites, for example, the ratio $K^{(i)}/K^{(i-1)}$ is successively $4k$, $3k/2$, $2k/3$, and $k/4$ (with the convention that $K^{(0)}$ is unity). There is apparently decreasing affinity for free ligand as more sites are filled on the macromolecule. This qualitatively appears to fit the definition of negative cooperativity, a decrease in binding affinity with increasing site occupation. However, here the appearance of negative cooperativity is deceptive. The decrease in overall affinity for ligands is due simply to the statistical effects of site occupation that progressively reduce the chance that a free ligand can find an unoccupied site on a partially saturated macromolecule. By the very construction of the model, *there is no interaction among the sites*, and hence no cooperativity, either positive or negative. Thus, this model represents *completely noncooperative binding*, and it can be used as a standard for comparison to detect site-site interactions.

4.3.2 THE ALL-OR-NONE MODEL AND THE HILL PLOT

The opposite extreme to the equal-and-independent sites model is the all-or-none model (see the bottom part of Figure 4.4). Suppose that the binding sites now interact so strongly that they have lost all their independence; they must act together as a single unit (a *cooperative unit*). Either none of the sites will accept a ligand, or else they all simultaneously take up a ligand each. Of course, the all-or-none model takes the notion of cooperativity to an extreme; real binding systems can only approach this degree of cooperativity.

Once again, the binding equilibrium for this model is

$$M + nL \underset{\longleftarrow}{\overset{K^{(n)}}{\longrightarrow}} ML_n \tag{4.8}$$

The isotherm equation for this model is

$$Y = \frac{<r>}{n} = \frac{K^{(n)}[L]^n}{1 + K^{(n)}[L]^n} \tag{4.9}$$

or alternatively

$$K^{(n)}[L]^n = \frac{Y}{1-Y} \tag{4.10}$$

One way to characterize the cooperativity in a binding system is to determine the number of cooperating units; for the all-or-none model this of course would be the parameter n. Now, the last equation can be recast in the suggestive form

$$\log\left(\frac{Y}{1-Y}\right) = \log K^{(n)} + n \log[L] \tag{4.11}$$

which neatly separates the two parameters n and $K^{(n)}$ and so indicates a route to their evaluation. This is the Hill plot, a graph of log $[Y/(1 - Y)]$ as a function of log [L], named after the English physiologist who suggested the all-or-none binding model and this equation as a way of describing the cooperative binding of oxygen to hemoglobin [18]. The Hill plot, though it has its weaknesses (see below), is frequently used to illustrate binding cooperativity.

If the binding system were to strictly follow the all-or-none model, then a Hill plot of the data should yield a straight line with a slope equal to the size of the cooperative unit, n. This slope is known as the Hill parameter or Hill coefficient, sometimes symbolized as n_H (some workers reserve n_H to indicate the slope at $Y = 0.5$). The x-axis intercept (where Y equals 0.5) can easily be manipulated to yield the apparent equilibrium constant $K^{(n)}$. Of course, for systems with a single site, the slope will be unity over the entire course of the plot.

Real binding systems with multiple sites and positive binding cooperativity will often have a Hill plot that is linear over a significant range of free ligand concentrations, typically centered approximately around the region for half saturation ($Y = 0.5$). In contrast, a Scatchard plot of the same binding data may have quite significant concave-down curvature over a major part of the plot. In these cases, the curvature in the Scatchard plot is certainly a good qualitative indicator of the presence of positive binding cooperativity, while the Hill plot is relatively less informative (see Figure 4.5). However, it is often difficult to quantitate such curved Scatchard plots by, e.g., curve fitting, to obtain estimates of the binding constants, the degree of cooperativity, and the stoichiometry. By comparison, it is relatively simple to obtain the Hill parameter and estimates of (at least some of) the binding affinities.

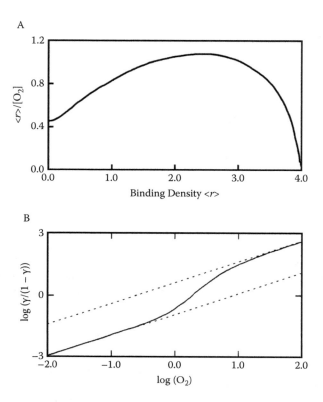

FIGURE 4.5 Comparison of Scatchard and Hill plots for the oxygenation of hemoglobin A. Binding curves calculated from the parameters given by Tyuma, I., Imai, K., and Shimizu, K., "Analysis of oxygen equilibrium of hemoglobin and control mechanism of organic phosphates," *Biochemistry* 12, 1491, 1973, for stripped hemoglobin A at 25°C, in 0.01 M Tris-HCl, pH 7.4. Hemoglobin tetramer concentration was 1.5×10^{-5} M. The isotherm equation used to fit experimental curves was

$$Y = \frac{K_1 p + 3K_1 K_2 p^2 + 3K_1 K_2 K_3 p^3 + K_1 K_2 K_3 K_4 p^4}{1 + 4K_1 p + 6K_1 K_2 p^2 + 4K_1 K_2 K_3 p^3 + K_1 K_2 K_3 K_4 p^4}$$

with oxygen partial pressure p measured in mm Hg and $K_1 = 0.114$ mm^{-1}, $K_2 = 0.165$ mm^{-1}, $K_3 = 1.17$ mm^{-1}, and $K_4 = 4.04$ mm^{-1}. A. Scatchard plot. The solid line represents a fit to the Adair isotherm with four affinity constants. B. Hill plot of the same data. Dashed lines represent asymptotes corresponding to values for the first and fourth affinity constants. Slope at half saturation (n_H) is approximately 2.5.

For example, the Hill plot permits easy estimation of the phenomenological binding constant $K^{(n)}$ of Eq. 4.11. For the all-or-none binding model the concentration of ligand needed to reach half saturation simply equals the reciprocal of $K^{(n)}$, for the uptake of n ligands by the cooperating sites. For real systems in which the behavior may be cooperative but not all or none, this half-saturating ligand concentration is at least a rough guide to the order of magnitude of the binding constants to be expected.

Provided that data extend to low or high enough degrees of binding saturation, the Hill plots for cooperative systems eventually exhibit a sigmoid character, indicating changes in cooperative interactions among the sites as they are filled. Wyman [19] has pointed out that in the limits of very low or very high degrees of saturation, the slope of a Hill plot will approach unity. The region of very low binding saturation corresponds to (noncooperative) uptake of a first ligand with occupation of a single site on the receptor. The limiting slope of unity at high binding saturation corresponds to uptake of a single ligand into the last available binding site.

In published Hill plots one often sees two asymptotes included along with the binding data. These two asymptotes, for high and low binding saturation, can be used to estimate the binding constants for both the first and the last ligand to bind. The necessary formulas can be obtained by analyzing the general expression for $Y/(1 - Y)$, Eq. 4.4 above, under limits in which the concentration of ligand is very small and very large:

$$\lim_{[L] \text{ and } Y \to 0} \frac{Y}{1-Y} = \frac{K^{(1)}[L]}{n} \tag{4.12}$$

and

$$\lim_{[L] \to \infty \text{ and } Y \to 1} \frac{Y}{1-Y} = n \left(\frac{K^{(n)}}{K^{(n-1)}} \right)[L] \tag{4.13}$$

From the limiting relations for $Y/(1 - Y)$ above it is seen that

$$\lim_{Y \to 0} \log \left(\frac{Y}{1-Y} \right) = \log \left(\frac{K^{(1)}}{n} \right) + \log[L] \tag{4.14}$$

where $K^{(1)} = nk$ is the phenomenological binding constant for binding the first ligand. Also,

$$\lim_{Y \to 1} \log \left(\frac{Y}{1-Y} \right) = \log \left(\frac{n K^{(n)}}{K^{(n-1)}} \right) + \log [L] \tag{4.15}$$

where $K^{(n)}$ is the phenomenological binding constant for binding n ligands simultaneously. The equilibrium constant for the addition of one more ligand to a system with $(n - 1)$ ligands already bound would be $K^{(n)}/K^{(n-1)}$, which could be interpreted as an effective binding constant for the last ligand to bind. This implies that if one extrapolates the asymptotes to intercepts where log $[Y/(1 - Y)]$ is zero, the corresponding values of log $[L]$ will give the logarithms of the (effective) binding constants for the first and last ligands to bind. There may be considerable experimental difficulty in making accurate measurements of Y in these regions, however, and so the determination of the asymptotes and their extrapolated intercepts may be subject to appreciable uncertainty.

The connection of the Hill plot to a phenomenological formulation of binding is straightforward but rather clumsy to express. For example, for a system with four binding sites and a phenomenological binding polynomial $\mathbf{P} = 1 + K^{(1)}[L] + K^{(2)}[L]^2 + K^{(3)}[L]^3 + K^{(4)}[L]^4$, one has

$$\log\left(\frac{Y}{1-Y}\right) = \log\left[\frac{K^{(1)}[L] + 2K^{(2)}[L]^2 + 3K^{(3)}[L]^3 + 4K^{(4)}[L]^4}{4 + 3K^{(1)}[L] + 2K^{(2)}[L]^2 + K^{(3)}[L]^3}\right] \tag{4.16}$$

From this it is clear that for this four-site case, and indeed for the more general case of an arbitrary number of sites, the Hill plot cannot be expected to be linear, except for a most unusual combination of binding affinities (e.g., the all-or-none model). Cornish-Bowden and Koshland [20] have analyzed the various possibilities for positive and negative cooperativity and the resulting curvature in the Hill plot.

Furthermore, it is algebraically possible to have the slope vary quite considerably (though it must always be positive), depending on the relative values of the phenomenological binding constants. The slope of the Hill plot is

$$\frac{d(\log[\frac{Y}{1-Y}])}{d\log[L]} = \frac{[L]}{Y}\left[\frac{1}{1-Y}\right]\frac{dY}{d[L]} \tag{4.17}$$

Evaluating this at the point of half saturation gives

$$n_H = \frac{[L]}{4}\frac{dY}{d[L]} = \frac{1}{4}\frac{dY}{d\ln[L]} \tag{4.18}$$

Upon substitution for the four-site binding polynomial from above, the slope is given by a ratio of polynomials in [L], including terms in powers of the free ligand concentration up to $[L]^{16}$. There is no a priori reason to expect that such a ratio of polynomials will collapse to a constant, for all values of [L] and an arbitrary set of binding affinities, and so one can (again) expect nonlinear Hill plots for arbitrary choices of affinities in multisite binding systems.

For general n-site systems, the Hill coefficient is not necessarily equal to the number of sites. The *maximum possible* value of the Hill coefficient can be shown to equal n, the number of sites, regardless of the particular binding model. For a real system, however, the slope at half saturation may not necessarily reach this maximal value, and it may in fact be considerably less. Consider, for example, noncooperative binding to n equal and independent sites. In this case the slope of the Hill plot will be unity everywhere because the number of interacting sites is just one, not n. Thus, in general the Hill parameter is not equal to the total number of sites.

Must the Hill coefficient be an integer? The algebraic analysis given above shows that it is, of course, quite possible to get Hill coefficients that are not integers. In fact, Hill found n_H to be 2.8 for the oxygen-hemoglobin system. Generally a noninteger value for n_H indicates the failure of the all-or-none model; intermediate states of binding, weighted by their respective population fractions, will produce nonintegral values for n_H . It can be shown that the Hill coefficient is a measure of the statistical variance of the binding population, that is, the standard deviation of the degree of saturation Y [19,21,22]. High positive cooperativity will tend to deplete intermediate states of ligation by comparison to the absence of cooperativity. This produces a higher degree of population dispersion than would happen with noninteracting sites and so gives a Hill coefficient greater than unity. Can the Hill coefficient ever be less than unity? Yes, that can happen too, and it can indicate either negative cooperativity among the sites (a collapse of the population toward intermediate states, more so than for noninteracting sites), or else two or more populations of noninteracting sites with different intrinsic binding affinities (or perhaps a mixture of the two cases). Unfortunately, the Hill plot is not well designed to distinguish among these possibilities. Section 4.3.4 elaborates on the mathematics of cooperativity and dispersion in the binding population.

4.3.3 An Operational Definition for Cooperativity

To move away from assumptions about the microscopic behavior of the system, it is possible to take a strictly operational approach and to define a quotient Q, an apparent equilibrium constant for ligand uptake, as

$$Q = \frac{Y}{[L](1-Y)} \tag{4.19}$$

(See Watari and Isogai [23] and Whitehead [24].) Assuming that the ligand L behaves ideally so that activity corrections can be ignored, Q will equal the binding constant for a system with either a single binding site or with n identical and independent binding sites. In either of these cases there is clearly no cooperativity present, and Q will remain a constant, regardless of the particular value of Y or [L]. Suppose, however, that Q increases with [L], at least for certain ranges of ligand concentration; then the derivative $dQ/d[L]$ will be positive in this range of [L]. The affinity of the receptor for the ligand is increasing as it takes up ligand, and for those values of [L] the system exhibits positive cooperativity. Also, the magnitude of the derivative $dQ/d[L]$ should indicate the magnitude of the cooperativity in some way. Conversely, where Q decreases with [L], the binding affinity is decreasing as more ligand is bound, and so here $dQ/d[L]$ will be negative; this is the range in which there is negative cooperativity, and again the magnitude of $dQ/d[L]$ should approximately indicate the degree of negativity. Finally, noncooperative binding occurs where $dQ/d[L]$ is zero. This suggests that one could monitor the derivative $dQ/d[L]$ as a function of [L] to determine the sign of the cooperativity and perhaps its magnitude.

It is important, however, to keep in mind that for a good determination of a binding isotherm the data should represent binding over a range of [L] that spans two or more orders of magnitude, in order to include binding at very low levels and binding that approaches saturation. Since it is difficult, in a simple linear plot, to adequately characterize the approach to saturation while still providing adequate separation of data at very low binding densities, Klotz has advocated the use of a semilogarithmic format for representing the isotherm [25]. Using a logarithmic scale for the ligand concentration will compress the data range and give an adequate view of binding at both extremes.

A similar logarithmic plotting scheme to follow log Q as a function of log [L] was suggested by Watari and Isogai [23]. This is a simple variation on the Hill plot, but one advantage of this format over that of the Hill plot is that it is easier here to detect cooperativity by eye. A system with no binding cooperativity (with equal independent sites) will yield a horizontal plot, not a slanted line of unit slope, as in the Hill plot. Humans have a difficult time discerning small slope deviations from a line at a 45° angle, so it is often difficult to see by eye whether the data in a Hill plot are consistent with positive or negative cooperativity. However, it seems to be much easier to detect slope deviations from the horizontal. With the plotting format suggested by Watari and Isogai, systems with cooperativity (negative or positive) will yield plot lines that are not horizontal. These systems are now easily distinguished by eye from noncooperative binding systems: a positive slope indicates positive cooperativity, while a negative slope indicates negative cooperativity.

Figure 4.6 compares some representative models and a simulation of a real experimental system, using the standard Hill plot and its modification suggested by

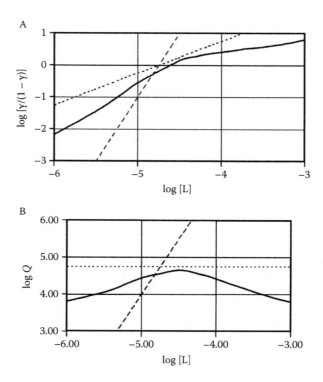

FIGURE 4.6 Detecting cooperativity: the Hill plot versus its modification by Watari and Isogai [23]. Simulations of four-site binding systems for NAD^+ binding to yeast glyceraldehyde 3-phosphate dehydrogenase. Solid line: Adair isotherm suggested by Cornish-Bowden and Koshland [20], with $K_1 = 4.6 \times 10^3$ M^{-1}, $K_2 = 1.4 \times 10^5$ M^{-1}, $K_3 = 7.5 \times 10^4$ M^{-1}, and $K_4 = 3.5 \times 10^3$ M^{-1}; dashed line: all-or-none binding model with k equal to 5.6×10^4 M^{-1}; dotted line: equal independent binding model with k equal to 5.6×10^4 M^{-1}. A. Hill plot. B. Watari-Isogai modification of the Hill plot.

Watari and Isogai (which will be called the W-I plot). The simulation is based on the isotherm equation proposed by Cornish-Bowden and Koshland [20] for the binding of NAD$^+$ to yeast glyceraldehyde 3-phosphate dehydrogenase, based on the data of Cook and Koshland [26]. Cornish-Bowden and Koshland noted how this system exhibited both positive and negative cooperativity in the uptake of NAD$^+$ by the enzyme. The enzyme is composed of four identical subunits, so that at saturation the enzyme would hold 4 NAD$^+$. The uptake is initially quite cooperative, but as saturation is approached, the affinity apparently drops off rather sharply. Cornish-Bowden and Koshland suggested that the data could be fit with a four-site binding equation of the Adair form:

$$Y = \frac{K_1[L] + 3K_1K_2[L]^2 + 3K_1K_2K_3[L]^3 + K_1K_2K_3K_4[L]^4}{1 + 4K_1[L] + 6K_1K_2[L]^2 + 4K_1K_2K_3[L]^3 + K_1K_2K_3K_4[L]^4} \tag{4.20}$$

with binding constants of $K_1 = 4.6 \times 10^3$ M^{-1}, $K_2 = 1.4 \times 10^5$ M^{-1}, $K_3 = 7.5 \times 10^4$ M^{-1}, and $K_4 = 3.5 \times 10^3$ M^{-1}; the numerical factors multiplying the binding constants are there for statistical reasons, since these binding constants are defined as "intrinsic," as follows:

$$K_1 = \frac{[EL]}{4\,[E][L]}, \quad K_2 = \frac{2\,[EL_2]}{3\,[EL][L]}, \quad K_3 = \frac{3\,[EL_3]}{2\,[EL_2][L]}, \quad K_4 = \frac{4\,[EL_4]}{[EL_3][L]} \tag{4.21}$$

The curvature of this system in the Hill plot is rather subtle, but the W-I format brings out the rise and fall in cooperativity quite nicely.

To compare this to the complete absence of cooperativity, Figure 4.6 shows a simulated binding curve for NAD$^+$ binding to an enzyme with four equal and independent binding sites, setting k equal to 5.6×10^4 M^{-1}, the average of the four binding constants in the Adair equation above. As expected, this gives a line of unit slope in the Hill plot, and a perfectly horizontal line at log $Q = $ log (5.6×10^4) in the W-I plot. To model extreme positive cooperativity, the figure has a simulation of a four-site all-or-none binding model, with $\mathbf{P} = 1 + (k\,[L])^4$ and k again set equal to 5.6×10^4 M^{-1}. This model yields straight lines of slope 4 in the Hill plot, but a slope of 3 (not 4!) in the W-I plot (it is easy to show that the n-site all-or-none model should give a slope of $n - 1$ for this latter format).

The Hill plot for this system (see Figure 4.6 A) initially shows a positive slope that approaches a value of 2, which indicates positive (homotropic) cooperativity. However, at about the titration midpoint ($Y \sim \frac{1}{2}$, with [NAD$^+$] $\sim 10^{-4}$ M) the slope starts to decrease, and the slope passes below unity, dropping to about 0.3 to 0.4. The cooperativity has now become distinctly negative. The W-I plot for this system (Figure 4.6B) shows this reversal of cooperativity rather more clearly. Apparently, the enzyme will avidly acquire NAD$^+$ at fairly low concentrations of the nicotinamide and then maintain a degree of saturation near 0.5 to 0.75, but will resist further uptake beyond that. The biological significance of this behavior is not completely clear, but it may have to do with buffering the levels of reduced versus oxidized nicotinamides in the cell.

4.3.4 Linkage Relations and Binding Cooperativity

The quantity Q defined above suffices to set a criterion for homotropic cooperativity, but what to do when binding of a second ligand species can influence the binding of the first? This directly involves binding linkage, and the binding polynomial approach is a powerful tool in dealing with such relations. It is time now to exploit some general properties of the binding polynomial in order to quantify both homotropic and heterotropic cooperativity.

Recall that a first differentiation of $\ln \mathbf{P}$ with respect to the logarithm of the concentration of ligand will yield $< r >$, the average number of ligands bound per macromolecule. This average takes account of the fraction of macromolecules with zero ligands bound, one ligand bound, two ligands bound, etc. In other words, it is the mean value of some underlying distribution of values of r. One might expect, then, to find that this distribution of r values has some spread to it. From the mathematical properties of the binding polynomial, it can in fact be shown that a second differentiation of $\ln \mathbf{P}$ gives information about the fluctuations in the number of ligands bound, that is, the statistical spread (the *variance*) in the binding density [19,21]. For systems with two or more species of ligand it is possible to obtain the *covariances* in the distribution of ligands as well.

For simplicity, a system with only two species of ligand will be treated. For such a system the variances σ_1^2, σ_2^2, and the covariance σ_{12}^2 for the distribution are defined by

$$\frac{\partial < r_1 >}{\partial \ln [L_1]} = < (r_1 - < r_1 >)^2) > = \sigma_1^2 \tag{4.22}$$

$$\frac{\partial < r_2 >}{\partial \ln [L_2]} = <(r_2 - < r_2 >)^2) > = \sigma_2^2 \tag{4.23}$$

$$\frac{\partial < r_1 >}{\partial \ln [L_2]} = \frac{\partial < r_2 >}{\partial \ln [L_1]} = <(r_1 - < r_1 >)(r_2 - < r_2 >) > = \sigma_{12}^2 \tag{4.24}$$

The angled brackets indicate averages taken over all possible states, while r_1 and r_2 denote particular numerical values of the binding density at a given ligand concentration. These quantities can be directly related to the Hill coefficient, and they provide a quantitative thermodynamic measure of the cooperativity in binding.

To quantitatively define homotropic cooperativity, first consider a system with n equivalent sites and a single ligand species L_1 that binds independently. The variance for this system is easily found to be

$$(\sigma_1^0)^2 = \frac{n\,k\,[L]}{(1 + k\,[L])^2} = nY_1(1 - Y_1) \tag{4.25}$$

The superscript 0 on $(\sigma_1^0)^2$ specially indicates that, by definition, cooperativity is absent for this system. Now suppose there is a system with the same number of sites, but in which there is interaction among the sites as ligands are bound. There will be a variance $(\sigma_1)^2$ in the binding density for this system that presumably is not the same as for the system with independent, equivalent binding. The Hill coefficient $n_{H,1}$ for this latter system may be shown to be [19,21,22]

$$n_{H,1} = \left(\frac{\partial\left(\frac{Y_1}{1-Y_1}\right)}{\partial \ln[L_1]} \right) = \frac{1}{nY_1(1-Y_1)}\sigma_1^2 = \frac{\sigma_1^2}{\left(\sigma_1^0\right)^2} \tag{4.26}$$

A similar relation, involving $(\sigma_2)^2/(\sigma^0_2)^2$, holds for $n_{H,2}$.

This relation is independent of any particular microscopic model for the system with site-site interactions, save for the specification of the number of sites n. The relation shows how the Hill coefficient compares the variance of the system with cooperative interactions to that expected for a system with no cooperativity at all. With positive cooperativity ($n_{H,1} > 1$) the system has a greater tendency to occupy states at the extremes of either low binding density (few ligands bound) or high binding density (more bound) than would the reference system with equal and independent site binding. Negative cooperativity ($n_{H,1} < 1$), on the other hand, corresponds to a system that tends to avoid extremes and to contract itself more toward the "middle" (the mean) than would the reference noncooperative system.

Heterotropic cooperativity will be defined here by following Schellman [21], taking as the reference system a macromolecule with two separate kinds of ligand and their respective binding sites, and assuming equal and independent binding for each ligand species over their binding sites. The binding polynomial here is

$$\mathbf{P} = (1 + k_1[L_1])^{n_1} \times (1 + k_2[L_2])^{n_2} \tag{4.27}$$

and the binding densities are

$$\left(\frac{\partial \ln \mathbf{P}}{\partial \ln[L_1]} \right)_{[L_2]} = <r_1> = \frac{n_1 k_1[L_1]}{1 + k_1[L_1]} \tag{4.28}$$

and

$$\left(\frac{\partial \ln \mathbf{P}}{\partial \ln[L_2]} \right)_{[L_1]} = <r_2> = \frac{n_2 k_2[L_2]}{1 + k_2[L_2]} \tag{4.29}$$

Now differentiating these relations a second time yields

$$\left(\sigma_1^0\right)^2 = n_1 Y_1(1 - Y_1), \quad \left(\sigma_2^0\right)^2 = n_2 Y_2(1 - Y_2), \quad \left(\sigma_{12}^0\right)^2 = 0 \tag{4.30}$$

A covariance $(\sigma_{12}^0)^2$ equal to zero indicates the complete independence of binding of species 1 from species 2. Automatically this gives a criterion for heterotropic cooperativity—a nonzero value for the covariance $(\sigma_{12})^2$ of the interacting system. A positive covariance indicates that binding of one species reinforces (makes more probable) some binding of the other; a negative value indicates interference, that is, binding of one species reduces the probability of binding of the other. As for quantifying homotropic cooperativity in the presence of multiple ligand species, one can use the ratio of variances, $(\sigma_1)^2/(\sigma^0_1)^2$ and $(\sigma_2)^2/(\sigma^0_2)^2$, essentially as before.

Notice that these variances and covariances are still functions of the concentrations of the ligand species; they can change across a titration. Thus, it is possible to have quite different degrees of cooperativity as a titration proceeds; it is even possible to reverse the sign of the cooperativity. This sign reversal appears in the example of the binding of NAD^+ to yeast glyceraldehyde 3-phosphate dehydrogenase that was discussed above.

One drawback to this approach to cooperativity analysis is that it calls for quite a bit of data to be collected, in order to evaluate the various statistical averages. However, the statistical analysis can be done without referring to any microscopic model of the binding, except for the specification of the number of sites. If a particular microscopic model is then chosen, the formulas above can be used to relate the statistical quantities to the particular parameters of the model.

4.4   STRUCTURAL MODELS OF COOPERATIVE BINDING

The Adair equation for the oxygenation of hemoglobin is a purely phenomenological description of the binding isotherm with a fourth-degree polynomial in dissolved oxygen concentration. The result of curve fitting shows that for hemoglobin, $K^{(4)}$ is hundreds of times greater than $K^{(1)}$, even after correction for statistical effects; this implies appreciable cooperativity in binding. However, no structural basis for the cooperativity is given at this level of description. In contrast to this, the Hill equation is based on a structural idea, that of n associated units acting together. Curve fitting of oxygenation data to the Hill equation is indeed consistent with the cooperation of several heme groups, since the Hill coefficient at half saturation is 2.8 [3,4]. Still, the Hill model of all-or-none binding is more phenomenological than structural, especially since it does not accommodate intermediate states of ligation between fully deoxygenated and fully oxygenated hemoglobin. Linus Pauling's 1935 models of pairwise interactions in tetrahedral or square-planar subunit arrays were really the first to incorporate structural ideas explicitly into the explanation of cooperativity [5]. Even then, the nature of the interaction between the heme groups was left undefined. Pauling referred to a "connection" between the heme groups, perhaps envisioning a direct electronic interaction, and it is clear that he envisioned the interaction as a short-range one.

It was known early on that crystals of deoxygenated hemoglobin would shatter upon exposure to oxygen [27]. This dramatic evidence of a change in protein shape and packing, together with the well-established Bohr effect for hemoglobin, provided the basis for the proposal by Wyman and coworkers for linking structural changes to ligand binding, culminating in the famous allosteric model of Monod, Wyman, and Changeux (MWC) [12]. The MWC model introduced a set of conformational changes in the protein and so allowed interactions between widely separated sites on the protein. The Koshland model of induced fit [28], as elaborated to encompass multi-subunit enzymes, appeared shortly thereafter (KNF) [13]. The KNF model used similar ideas, and in fact both the KNF and MWC models are special cases of a more general model, as described by Hammes and Wu [29].

The MWC model has been enormously influential, with applications in enzymology, pharmacology, physiology, and molecular biology. Both the MWC and the KNF models were developed to explain the kinetic behavior of enzymes, in particular the influence of small nonsubstrate molecules on the kinetics, and the frequent observation of non-Michaelis-Menten behavior tied to positive cooperativity. Since these models were developed to deal with enzymes, the terminology and concepts apply most appropriately to protein systems with interacting subunits and binding sites; a different set of models is needed to treat lattice systems (e.g, drug binding to DNA), and these other models will be discussed in Chapter 5.

4.4.1 THE CONCERTED MONOD-WYMAN-CHANGEUX (MWC) MODEL

The MWC model assumes that the receptor is composed of a relatively small number of identical subunits; the nature of the subunits is purposefully left undefined so as to retain generality in the theory. A subunit in an enzyme, for example, might be an individual polypeptide, a group of polypeptides, or perhaps just a domain in a single polypeptide chain. The identical subunits in the receptor are termed *protomers*, while *monomer* refers either to a fully dissociated protomer or to a group of nonidentical subunits in an identifiable assemblage, with the receptor made up of several of these assemblages.

The protomers in the receptor occupy equivalent positions. Thus, the overall receptor must have at least one axis of symmetry. For a given species of ligand, each protomer has one and only one site for binding it specifically. The protomers can reversibly switch between (at least) two different conformations that differ in the distribution and/or energy of interaction with neighboring protomers, and conformational switching is constrained by interaction with neighboring protomers. These different conformational states also differ in their affinity toward one or more of the ligand species that may be bound specifically. Lastly, when the receptor changes conformation, it does so in such a way as to maintain its molecular symmetry. This means that all the subunits in the receptor must change conformation together, hence the description of the model as "concerted." It also has the effect of simplifying the algebra involved in deriving the binding isotherm equation.

The basic features of the MWC model can be visualized by examining a dimeric receptor (Figure 4.7). Each subunit has two available conformations, here denoted by circles and squares.

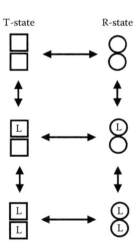

FIGURE 4.7 Six states of a dimeric receptor in the Monod-Wyman-Changeux model [12], with two available conformations, R and T, for the subunits. The assumption of a concerted conformational change restricts the total number of states, ligated or not, to six.

The traditional terminology for the two conformational states is T and R, for taut and relaxed. The R state has higher affinity for a primary ligand species than does the T state. The subunits can also bind a second, third, ... ligand species specifically as well, but the R versus T designation pertains only to the binding of the primary ligand.

Because of the requirement for a concerted conformational change, the two subunits must both be in the R state or both be in the T state; no mixed R/T receptor states are allowed in the model. There are six distinguishable states of ligation, as shown in the figure.

Taking $[T_0]$ as the concentration of unligated receptor in the T state and $[R_0]$ as the concentration of unligated receptor in the R state, the equilibrium constant for changing subunit conformation will be denoted as $K_{conf} = [T_0]/[R_0]$. (Historically, this equilibrium constant was denoted by the symbol L, but to avoid confusion K_{conf} will be used instead.) It is conventional to suppose that the subunits have identical affinities for the ligand, and that this affinity will change with subunit conformational state; k_R and k_T will be used here for the two different microscopic binding constants. Taking account of the conformational change, the binding polynomial for this model can be written as

$$\mathbf{P} = 1 + 2k_R[L] + k_R^2[L]^2 + K_{conf}(1 + 2k_T[L] + k_T^2[L]^2)$$
$$= (1 + k_R[L])^2 + K_{conf}(1 + k_T[L])^2 \tag{4.31}$$

and the binding equation is found to be

$$<r> = \frac{\partial \ln \mathbf{P}}{\partial \ln [L]} = \frac{2k_R[L] + 2k_R^2[L]^2 + K_{conf}(2k_T[L] + 2k_T^2[L]^2)}{1 + 2k_R[L] + k_R^2[L]^2 + K_{conf}(1 + 2k_T[L] + k_T^2[L]^2)} \tag{4.32}$$

or collecting terms and expressing the equation in terms of the degree of saturation Y

$$Y = \frac{k_R[L](1 + k_R[L]) + K_{conf}k_T[L](1 + k_T[L])}{(1 + k_R[L])^2 + K_{conf}(1 + k_T[L])^2} \tag{4.33}$$

The binding polynomial can of course be generalized for the case of a receptor with n identical subunits:

$$\mathbf{P} = (1 + k_R[L])^n + K_{conf}(1 + k_T[L])^n \tag{4.34}$$

and the corresponding binding equation follows easily.

In a direct plot of Y as a function of $[L]$, will this model binding curve be sigmoid and show cooperative binding? This depends on the particular values of the parameters k_R, k_T, and K_{conf}, but for a wide range of these parameters the model does indeed yield a sigmoid binding curve. By simplifying the model it is possible to explore how cooperativity is manifested for certain limiting cases.

First, suppose only the R state can bind ligand (this is the case of *exclusive binding*). Then k_T can be set equal to zero, and for a receptor with n subunits the binding equation becomes

$$Y = \frac{k_R[L](1 + k_R[L])^{n-1}}{(1 + k_R[L])^n + K_{conf}} \qquad (4.35)$$

Compare this to the equation for a system with n identical binding sites, but no conformational change:

$$Y = \frac{k_R[L](1 + k_R[L])^{n-1}}{(1 + k_R[L])^n} \qquad (4.36)$$

The difference lies in the extra term K_{conf} in the denominator. Conformational switching will reduce the number of available sites and so reduce the degree of saturation achievable at a given concentration of free ligand. If K_{conf} is small, that is, if the receptor population is mostly in the all-R conformation already when ligand is added to the system, then there will be little cooperativity apparent. If, on the other hand, K_{conf} is appreciably larger than unity, then a substantial fraction of the receptors will be in the all-T state at the beginning of the titration, while very few will be in the all-R state initially. Thus, the initial region of the binding isotherm will show only a slight rise in binding as ligand is added. At some point, however (depending on the value of K_{conf} relative to $k_R[L]$), receptors will gain sufficient free energy from taking up n ligands that they switch conformation from all-T to all-R. This causes a sudden apparent rise in the affinity for free ligand, and the isotherm curve climbs steeply to the saturation plateau. Cooperativity has appeared.

Allowing the T state to bind ligand also (this is called the *nonexclusive* model) will of course modify this picture and perhaps reduce the apparent degree of cooperativity. In any event, the T-state receptor's affinity for ligand is by assumption less than that for a receptor in the R state. If the affinities were the same in both states ($k_R = k_T$), then cooperativity disappears and this is just a system of n equal sites again, with, of course, a standard hyperbolic binding curve. The distribution of the receptor population over two conformations would not matter here.

4.4.1.1 Heterotropic Effectors in the MWC Model

So far, this serves as an explanation of homotropic cooperativity, of how the binding of one ligand of a given species can influence the binding of other ligands of the same species. Now on to heterotropic effects in the MWC model, in which binding of the second, different species L_2 can influence the binding of the first species L_1. Suppose that this second ligand species prefers the T conformation, so that $k_{2,R} < k_{2,T}$. Uptake of this second species by the receptors will drive them toward the T conformation, but the extent of the shift will depend on the value of K_{conf} relative to $k_{2,T}[L_2]$. Consider the case in which the presence of L_2 produces a considerable shift in the receptor population toward the all-T state (the symmetry requirement in the model will not permit mixed conformational states even though this is a different ligand species). Then this is about equivalent to the exclusive binding case discussed above, in which K_{conf} favored a receptor population in the all-T state. Once again, the binding curve for species 1 will be flat initially, but then it will rise sharply when $[L_1]$ is high enough to drive the conversion of receptors to the all-R state. So there is heterotropic cooperativity when species 2 acts as an allosteric inhibitor of the binding of species 1.

Conversely, if species 2 were to bind preferentially to the R conformation, then increasing $[L_2]$ would shift the receptor population toward the all-R state. This would promote the binding of L_1; species 2 could be considered an allosteric activator of the receptor, with respect to uptake of species 1. The sigmoidicity of the binding curve for species 1 would be decreased, however, so that the cooperativity in binding species 1 would apparently be less.

4.4.2 THE SEQUENTIAL KOSHLAND-NEMETHY-FILMER (KNF) MODEL

The KNF model is based on Koshland's model of *induced fit* for enzyme action. The theory of induced fit was developed to explain why some enzymes showed a lack of reactivity toward potential substrates. According to the lock-and-key model of enzyme action, a substrate would fit into the active site of an enzyme as a key would fit in a lock; the enzyme's active site is a sort of negative or complementary image of the substrate. The lock-and-key model explains much of the extraordinary specificity shown by enzymes in catalyzing reactions, but it fails in certain cases, notably those in which it would predict that water would be a substrate and experiment has found that it definitely is not. Koshland [28] proposed that in such cases the enzyme's active site would have a certain flexibility, allowing it to adapt its shape to match that of the substrate. The conformational change would be driven by favorable interactions between functional groups on the enzyme and the substrate. If there were not enough of these favorable interactions, then the conformational change would be incomplete or absent, and no catalysis would occur. The lack of reactivity of water in the active site would thus be explained as a lack of enough favorable contacts to drive the enzyme into a catalytically active conformation.

In 1966 Koshland, Nemethy, and Filmer applied the idea of induced fit to explain cooperativity and the effect of ligand binding on enzymatic activity [13]. As in the MWC model of multisubunit enzymes, the KNF model includes conformational changes. In the absence of ligand the (identical) subunits of the receptor exist in one conformation. Binding of a ligand to a subunit induces a conformational change in that subunit, and this change may be propagated to vacant neighboring subunits or sites (another similarity to the MWC model). However, unlike the MWC model, in the KNF model the subunits within a given receptor are *not* required to all be in the same conformation (see Figure 4.8 for application to a

FIGURE 4.8 Three states for a dimeric receptor in the sequential model of Koshland, Nemethy, and Filmer [13]. Binding of ligand L induces a conformational change; two different subunit conformations may coexist in the same receptor molecule.

dimeric receptor), and so the KNF model specifically allows for mixes of conformational states in the same receptor. The symmetry requirement of the MWC model does not appear here, either.

Since the subunits take up ligand and change conformation in sequence, the KNF model is also known as the *sequential* model. The basic model allows for only two subunit conformations, as shown, but it is possible to generalize the theory to allow for a third (fourth, etc.) conformation for unoccupied subunits that are in contact with one or more occupied subunits.

The binding polynomial for a dimeric receptor in the basic model must account for only three states. Using a single microscopic binding constant k for ligand binding and a cooperativity factor of α for stabilizing the doubly ligated state, the binding polynomial $\mathbf{P}$ can be written as

$$\mathbf{P} = 1 + 2\,k\,[\mathrm{L}] + \alpha\,k^2\,[\mathrm{L}]^2 \tag{4.37}$$

The factor of 2 in the second term on the right accounts for two alternative microscopic states of the singly ligated receptor. The α parameter represents interaction between neighboring (occupied) subunits; the idea of such a (stabilizing) interaction goes back to Pauling's 1935 model for hemoglobin. Positive homotropic cooperativity appears when α is greater than 1, negative cooperativity when α is less than 1. A major difference from the original MWC model is now apparent: the KNF model can accommodate negative cooperativity, but the MWC model (at least in its basic formulation) could not.

4.4.2.1 Heterotropic Effectors in the KNF Model

To see how heterotropic effectors within the KNF model might be included, consider again the case of a dimeric receptor with identical subunits and two different subunit conformations. Suppose that the second ligand species (denoted X), like the first species (denoted L), will cause a subunit to switch conformation, but that when bound to a subunit the species X does not directly compete with L for binding. Ten distinguishable states of ligation for the receptor are present, with up to six different cooperative interactions between subunits, as shown in panel A of Figure 4.9.

Depending on the particular values of the different cooperativity parameters, ligand species X can act to promote or to hinder binding of species L; that is, one can have either positive or negative cooperativity. Furthermore, the cooperativity can change throughout the titration, for example, going from positive to negative, and back to positive again. This flexibility in incorporating cooperative interactions into the model comes at the price of increased complexity in the model, since there are now two different binding constants and up to six different cooperativity parameters.

Matters can be simplified by supposing that ligand species X acts as a competitor to the primary ligand in binding. There are then only six states and three cooperativity parameters, as shown in panel B of Figure 4.9. Other models can also be easily constructed along these lines, to allow for induction of a third subunit conformation, etc.

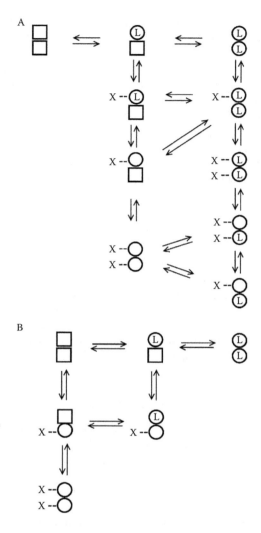

FIGURE 4.9 Heterotropic cooperativity in a KNF dimer model. Both the ligand L and the effector X induce a change from the T to the R form. A. X does not compete with L in binding to subunits; there are 10 possible states for the dimer. B. X competes directly with L; there are now six possible states for the dimer.

4.4.3 COMPARISON OF THE KNF AND MWC MODELS

Both the basic KNF model and the basic MWC model suppose that the receptor is a symmetric arrangement of identical subunits, that the subunits can change conformation, and the conformations differ in their affinity for ligand. The switching of conformation is tied very closely to the origin of cooperativity, so that it is not just a matter of having sites with different affinity. Cooperativity in both models arises as the population of receptors shifts in conformation and hence in affinity.

The MWC model is especially restrictive with respect to this conformational switch: the subunits in a receptor must switch conformation in concert, and the overall symmetry of the subunit must be maintained. The symmetry requirement in particular seems very artificial, and the KNF sequential model appears to be more realistic here. A further basic difficulty with the MWC model is that it makes no allowance for negative cooperativity, yet there are several well-known multisubunit enzyme systems that show negative cooperativity (reviewed by Levitzki and Koshland [30]). The KNF approach, on the other hand, easily incorporates both negative and positive cooperativity in the model.

The MWC and KNF models can be regarded as special cases of a more general model. For a tetrameric receptor with subunits arranged in a square, the different states of the general model can be written in the form of a 4 × 4 array, as shown in Figure 4.10. The sequential KNF model is composed of just the states lying on the principal diagonal indicated by the dotted line, while the states for the MWC model are the two columns on the left and right sides of the array. To describe these additional states, one would have to introduce more adjustable parameters (binding constants and cooperativity factors), which the quality and quantity of the collected data may not justify. The MWC and KNF models, at least in their simplest versions,

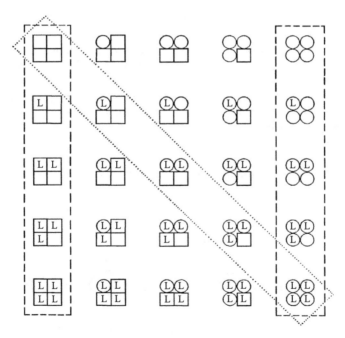

FIGURE 4.10 A generalization of the MWC and KNF models for a tetrameric receptor. The MWC model is represented by the states enclosed by the dashed lines, and the KNF model is represented by the states enclosed by the dotted lines. Free ligand L and the arrows joining the states have been omitted for clarity. Adapted, with permission, from *Science* Vol. 172, G.G. Hammes and C.-W. Wu, "Regulation of enzyme activity," pages 1205–1211. Copyright 1971 American Association for the Advancement of Science.

have the advantage of describing a wide range of behaviors of the titration curve while using only a small set of parameters.

4.4.4 OXYGENATION OF HEMOGLOBIN

A natural application of these models would be to the oxygenation of hemoglobin, with its four subunits that bind oxygen cooperatively. There are three simple ways to arrange the four subunits: a linear arrangement, a square planar arrangement, and a tetrahedral arrangement of the subunits. As noted before, the square planar version has some historical significance, and it is fairly simple to treat, though it is now known that the subunits are arranged tetrahedrally and that the binding cooperativity is more subtle and more complicated than in this simple model.

Suppose that binding induces a conformational change from the T state to the R state, with a microscopic site binding constant of k for each of the subunits; this is in effect an "induced fit" model. In Figure 4.11 the two subunit conformations are indicated with squares (for the T state) and circles (for the R state, with oxygen bound to the heme group). Now it is possible to distinguish six different states for the receptor. Notice that there are two different states possible when the tetramer has bound two ligands. One also must take account of the number of ways of arranging ligands for a particular state. For example, there will be four ways to place a single ligand onto an unoccupied tetrameric receptor, but only one way to add a ligand to a receptor with three ligands already bound.

Furthermore, it is necessary to allow for pairwise stabilizing interactions among the subunits. These interactions are indicated in the figure by the dashed lines joining the subunits. Notice that when two ligands are bound, in one of the arrangements there will be no stabilizing interaction, but in the other arrangement, two of the subunits may interact and enjoy an extra contribution to the statistical weight of α for this state. (The use here of the symbol α for cooperativity should not be confused with the conventional designation of one of the two types of polypeptide chains in the hemoglobin tetramer.) With three ligands bound, there are two such stabilizing interactions and a contribution of α^2 to the statistical weight, and when four are

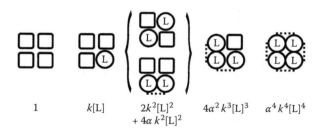

FIGURE 4.11 Simple square planar model for hemoglobin oxygenation. Four identical subunits follow an induced-fit mechanism and switch between taut (small square) and relaxed (circle) conformations, with multiple states of ligation. Dashed lines joining subunits indicate cooperative interactions between them.

bound, there are four interactions and a contribution of α^4. This results in a binding polynomial of the form

$$\mathbf{P} = 1 + 4k[L] + 4k^2\alpha[L]^2 + 2k^2[L]^2 + 4k^3\alpha^2[L]^3 + k^4\alpha^4[L]^4 \qquad (4.38)$$

which is essentially that proposed by Pauling in 1935, and which gave a reasonably satisfactory fit to the data available at that time. This allows for contributions from all states of oxygenation of the hemoglobin tetramer, and in particular it permits appreciable amounts of intermediate states of oxygenation, that is, doubly and triply ligated hemoglobin.

Work by Stanley Gill and coworkers [31] has shown, however, that in fact the triply ligated state of hemoglobin is scarcely populated, so that the statistical weight for this state should be reduced considerably from that assigned in the above model. To account for this and other observations on the order of subunit binding, Di Cera, Robert, and Gill proposed a model for hemoglobin oxygenation that reflected the heterogeneous subunit composition of hemoglobin, its subunit stoichiometry of $\alpha_2\beta_2$, and the tetrahedral arrangement of subunit contacts (Figure 4.12) [32]. In their model

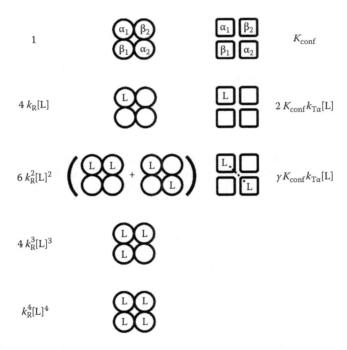

FIGURE 4.12 Model for hemoglobin oxygenation proposed by Di Cera et al. [32]. While the molecule is in the T state, oxygen binding to the β subunits is not allowed, hence there are no triply or quadruply ligated T state molecules shown. The statistical weights shown on the margins reflect the assumptions that while the molecule is in the R state, the binding is not cooperative; that binding affinities to both α and β subunits are the same for molecules in the R state, that cooperativity in binding to the T state can be parameterized with a single quantity γ, and that a single equilibrium constant K_{conf} governs the switching between T and R states.

the hemoglobin would exist in two quaternary states, T and R, with an equilibrium constant K_{conf} for the conformational change, and no mixed states would be permitted (just as with the MWC model). While the tetramer was in the T state, oxygen would be bound only by the α subunits (binding constant $k_{T,\alpha}$), with cooperative interactions allowed between the two α subunits; there would be no binding at the β subunits for tetramers in the T state. The parameter γ is introduced to account for the cooperativity: γ could be greater than unity, for positive cooperative interaction between the α subunits; or it might be less than unity, for negative cooperativity; and if it were equal to unity, then there would be no cooperativity at this level. Subunits in the R state would have no binding cooperativity, but now the β subunits would have equal affinity with the α subunits (both with binding constant $k_R > k_{T,\alpha}$). In this way the triply ligated state is de-emphasized since it is permitted only for the R state and not the T state. At the same time, cooperativity is maintained through the parameter γ and the all-or-none switching to the R state.

The binding polynomial for this model would be

$$\mathbf{P} = (1 + k_R x)^4 + K_{conf}(1 + 2k_{T\alpha}x + \gamma k_{T\alpha}^2 x^2) \tag{4.39}$$

where x is the ligand activity. This differs in formulation from that used by Di Cera et al. by a factor of $([T_0] + [R_0])/[T_0]$; under their convention, but not ours, the binding polynomial is scaled by the fraction of the population in the R state. This should have no effect on the binding density, however. The model gives a good fit to the oxygenation curve, appreciably better than that afforded by a strict MWC model.

The Ackers group has gone on to develop an even more detailed model of hemoglobin binding cooperativity, breaking out the overall oxygenation process into 10 microstates and assigning stepwise cooperative free energy values for each state. This model is based on recent structural studies showing multiple conformations for ligated hemoglobin [33]. The structural studies have been complemented by very cleverly designed experiments using a variety of altered hemoglobins to track ligation and conformational changes in intermediate states of oxygenation. These include hemoglobins with mutated subunits (e.g., hemoglobin S, with the altered β subunit); hemoglobins with cobalt, zinc, or manganese replacing the ferrous iron ion in the heme group; and hemoglobins ligated to molecules such as CO or CN$^-$, to block subunit conformational changes and lock in the arrangement of ligands over the subunits (summarized in Ackers and Holt [34]).

Figure 4.13A shows these 10 microstates and their respective statistical weights in the binding polynomial. The quantity $^{ij}k_c$ is a cooperative interaction parameter for forming microstate (ij) from the unligated tetramer, and k_{int} is an intrinsic affinity constant for the binding of O_2 to a free hemoglobin $\alpha\beta$ dimer; there is no cooperativity in the oxygenation of the free dimer, so that this state provides a convenient thermodynamic reference point. Factors of 2 account for the two structurally distinct but functionally equivalent configurations within particular microstates. Figure 4.13B shows the experimental hemoglobin oxygenation cascade and the stepwise change in binding constants as oxygenation proceeds through the various possible intermediate microstates.

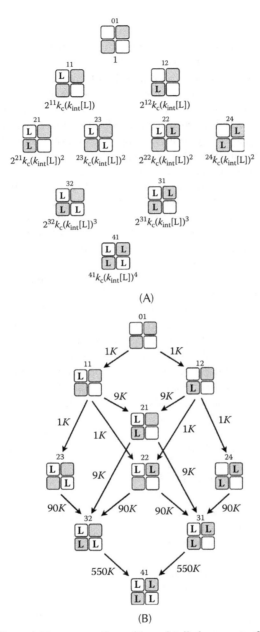

(A)

(B)

FIGURE 4.13 Hemoglobin oxygenation with a detailed account of binding cooperativity. Shading distinguishes α and β subunits. A. Microstates and their statistical weights. B. Ligation (oxygenation) cascade for hemoglobin, showing transitions among the 10 microstates of panel A, and the (approximate) relative augmentation of oxygen affinity (as multiples of the fundamental affinity K) for each transition. Adapted from *Journal of Biological Chemistry* Vol. 281, G.K. Ackers and J.M. Holt, "Asymmetric cooperativity in a symmetric tetramer: Human hemoglobin," pages 11441–11443, with permission. Copyright 2006 The American Society for Biochemistry and Molecular Biology.

4.4.5 NESTING

An interesting extension of the basic allosteric model is the concept of *nesting* [35]. Nesting provides for a second, third, or even higher level of allosteric regulation of binding and activity. It is based on grouping subunits together at a primary level and allowing allosteric interactions among them, as in the basic MWC or KNF models, then introducing another (higher) level of allosteric interactions between these primary groups of sites. There is good structural evidence of just such separate levels of allosterism in some systems, e.g., with arthropod hemocyanins [35,36] and with the oligomeric chaperonin GroEL of *E. coli* [37].

Obviously, a wide range of nesting schemes present themselves. One could nest MWC inside MWC, MWC inside KNF, or KNF inside KNF. Additionally, one could use models other than the MWC or KNF schemes for cooperativity within a level, or one could allow for subunit dissociation/association equilibria, etc.

For example, suppose one has an eight-subunit protein with eight identical polypeptide chains, and that these chains are assembled first as a ring of four, then grouped as a dimer of tetrameric rings. Also suppose that within each tetrameric ring the identical subunits follow the MWC scheme for maintaining symmetry during conformational change; that is, a ring exists either as an all-T form (represented as a block of cubes in Figure 4.14) or as an all-R form (represented by the ring of four circles in the figure). The T form dominates in the absence of ligand. Now suppose that at the level of the dimer the symmetry can be broken; either or both of the rings can switch conformation. As ligand is added the equilibrium shifts from the symmetrical TT state to the unsymmetrical TR (or RT) state (see the figure). With yet more ligand added, the equilibrium shifts toward the symmetrical RR state. This scheme embeds the MWC model for the individual rings within a grosser KNF model for the dimer of rings.

A binding polynomial for this model can be developed starting with the usual R and T designations for the two different conformational states in the tetrameric rings. For a single tetrameric ring in the T state the binding polynomial is $\mathbf{P}_T = (1 + k_T [L])^4$, and for a single ring in the R state it is $\mathbf{P}_R = (1 + k_R [L])^4$. (For simplicity all sites will be treated as equivalent within a ring without differentiating among geometrically distinguishable arrangements of ligands over a ring.) For a dimer of two tetramers the overall binding polynomial is $\mathbf{P} = \mathbf{P}_T^2 + 2 K_{RT} \mathbf{P}_T \cdot \mathbf{P}_R + \alpha K_{RT}^2 \mathbf{P}_R^2$, where the parameter α allows for cooperativity between rings, in converting the second ring from the T to the R conformation.

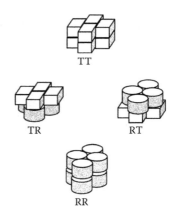

FIGURE 4.14 Nesting for a dimer of hexamers. Hexamers individually follow the MWC model of all-or-none conformational change, while at the level of the dimer the KNF model of sequential conformational change is followed.

Nesting of models has not been widely applied in interpreting experiments, but it has found application in certain large and complicated binding systems. The nesting concept has been applied, for example, to describe the unusual oxygenation patterns of arthropod hemocyanins where there are 24 sites for oxygen binding per protein molecule [36]. The data were fit using as a model a dimer of dodecamer units. The dimer was taken to have two overall quaternary states, R and T, which interconverted as in the standard MWC model. The units of the individual dodecamer assemblies were also supposed to interconvert between two conformational states, denoted r and t, so that overall there were four substates to distinguish: Rr and Rt, nested within the R state, and Tr and Tt, nested within the T state. Another recent application of nesting has been to the ATPase activity of the protein chaperonin GroEL from *E. coli* [37]. This protein is a dimer of two heptameric rings of identical subunits. The MWC model was used to describe the T/R equilibrium within each ring, and the KNF model used to allow for interactions between the rings (e.g., transitions between the TT, TR, and RR states for the dimer of rings).

4.5 AGGREGATION AND COOPERATIVITY

Aggregation of the receptor and/or of the ligand can lead to binding cooperativity. Standard assay conditions typically use concentrations of receptor or ligand, and conditions of pH, salt, and temperature, which are nonphysiological and which may promote dissociation of complexes that would be stable under conditions in vivo. Perhaps the conformation and hence the activity of the subunits depends on their state of aggregation. Or perhaps a binding or catalytic site is formed at the interface of one or more subunits, and the activity is disrupted by the dissociation of the subunits. Alternatively, there may be aggregational equilibria involving the ligand species, with one or another aggregated form being the preferred binding companion to the enzyme or receptor. Such aggregational equilibria, involving either ligand or receptor, can produce cooperativity without requiring any allosteric conformational changes at binding sites, and this cooperativity may be either positive or negative. This appears to be significant for the catalytic action of a number of enzymes [38]; also, aggregation of receptors is a well-known phenomenon in the immune system [39], and it appears to be important for membrane-bound hormone receptors as well [40,41].

There are a number of reasons that receptor systems may have evolved to use multiple subunits and aggregation as a means of regulation. First, aggregation permits the transfer of "information" from one subunit to another about the state of the assembly, through propagated conformational changes. Second, by requiring precise alignment of contacts between distinct subunits in forming a multisubunit assembly, the system avoids incorporation of defective subunits (such as might arise due to mistakes made in receptor biosynthesis), but at the same time avoids the burden of requiring (nearly) perfect biosynthesis of very lengthy polypeptide chains. Third, by reducing the exposed surface area there is less chance of inappropriate association with other proteins (not something to be ignored in the crowded interior of the typical cell). For enzymes, a fourth consideration is that aggregation can promote efficiency in catalysis by bringing into close spatial proximity some subunits that are functionally distinct; reaction intermediates may be passed directly from one catalytic unit to the next in a stepwise fashion without the dispersive release of reaction intermediates.

There have appeared over the years a number of theoretical mechanisms for the regulation of enzymatic activity through aggregational equilibria. In 1967 Frieden presented a treatment of reversible association-dissociation of enzyme subunits and of the kinetic effects that could be expected [42]. He examined the case of a monomer-dimer equilibrium, in which the monomer and dimer had different kinetic properties. This model would show normal (i.e., Michaelis-Menten-like, or hyperbolic) enzyme kinetics at low total enzyme levels, in which the action of free monomer would dominate the kinetics. At high total enzyme levels, again the behavior would be normal, but here the dimer's kinetic properties would dominate. At intermediate levels of enzyme the mix of monomer and dimer activity could produce a sigmoidal activity-versus-substrate curve, which would then appear as positive cooperativity.

At about the same time, Nichol et al. [43] and Kurganov [44] independently suggested models for cooperativity arising from equilibrium aggregation of enzyme subunits. For all these models the kinetic properties depend on the total concentration of enzyme, and the cooperative effect, if present, could disappear at the enzyme levels normally used for assays, yet could play an important role under the quite different conditions obtained inside a cell.

Another strategy for regulating enzyme or receptor activity or specificity would be to use a piggy-back ligand. As discussed in Chapter 3, a piggy-back ligand is a secondary ligand species that produces its effects by binding to the main ligand species, and not directly to the receptor; this distinguishes it from the classical allosteric effectors of the MWC or KNF models of cooperativity. Yet another strategy would be to exploit any propensity for the ligand itself to aggregate, to create cooperativity. This section will first look at the effects of ligand oligomerization, then at the effects of receptor oligomerization, and finally it will briefly consider a model piggy-back system in which the secondary ligand drives aggregation of the primary ligand.

4.5.1 Aggregation of Ligand as a Source of Binding Cooperativity

For simplicity the discussion will be limited here to the effects of a ligand dimerization preequilibrium on the binding curve. Suppose that the receptor has two sites, a and b, in close spatial proximity; a and b are not necessarily identical, but they can individually bind the ligand as separated monomers, or they can simultaneously act together to bind the dimer. Four states for the receptor can be distinguished: vacant, the a site only occupied, the b site only occupied, and lastly, both a and b occupied by a ligand dimer. Dimer binding to either the a or b site will not be allowed. The dimerization of the ligand is governed by

$$L + L \xrightleftharpoons{k_{dimer}} D, \quad [D] = k_{dimer}[L]^2 \qquad (4.40)$$

(the ligand dimer here is denoted by D). The binding polynomial is then

$$\mathbf{P} = 1 + k_a[\text{L}] + k_b[\text{L}] + k_D[\text{D}]$$
$$= 1 + (k_a + k_b)[\text{L}] + k_D k_{\text{dimer}}[\text{L}]^2 \tag{4.41}$$

If instead of dimer binding there were binding of monomeric ligand to independent sites a and b, the last term in $\mathbf{P}$ would be replaced by $k_a k_b [\text{L}]^2$. To get a noticeable difference in the binding curve due to dimerization, there must be an appreciable difference between $k_a k_b$ and $k_D k_{\text{dimer}}$. In other words, the ratio $\omega = k_D k_{\text{dimer}}/k_a k_b$ ought to be significantly different from unity. This ratio is essentially a measure of the cooperativity of dimer binding over independent binding of monomers. In terms of free energy changes, this is

$$\Delta G_{coop} = -RT \ln \omega = -[RT \ln(k_D k_{\text{dimer}}) - RT \ln(k_a k_b)] \tag{4.42}$$

A zero free energy change here means no net effect of dimerization on the binding to the receptor, while negative values of ΔG_{coop} of course will favor dimer binding.

With ΔG_{coop} equal to 0 so that $k_D k_{\text{dimer}} = k_a k_b$, the system resolves to one with two independent binding sites with different binding affinities. With k_D set equal to $k_a k_b$, cooperative behavior appears as the dimerization constant is increased from 1 to 100 M^{-1}. Figure 4.15 compares isotherms in two different plotting formats for this system, and shows that positive cooperativity in this system is perhaps more apparent when the data is plotted in the Scatchard format than as a direct plot. In the Scatchard format, in the absence of dimerization cooperativity (for $k_{\text{dimer}} = 1$) the curve is concave up, with an intercept on the x-axis at $n = 2$, and an intercept on the y-axis at $<r>/[\text{L}] = k_a + k_b$. As the dimerization equilibrium constant increases, the intercepts remain the same but the curvature reverses; that is, the characteristic "hump" for positive binding cooperativity begins to appear in the plot. A modest degree of positive cooperativity is apparent when the dimerization constant is only 10 M^{-1}, and it becomes quite pronounced for $k_{\text{dimer}} = 30$ M^{-1}. By comparison, the direct plot shows rather little evidence of any sigmoid behavior that would indicate positive cooperativity.

It may be that the binding of individual monomers is quite negligible compared to dimer binding. Neglecting the binding of monomers (so that now only the dimer can bind) will give

$$\mathbf{P} = 1 + k_D[\text{D}] = 1 + k_D k_{\text{dimer}}[\text{L}]^2 \tag{4.43}$$

and the binding equation is (in terms of free monomer concentration)

$$<r> = \frac{2 k_D k_{\text{dimer}}[\text{L}]^2}{1 + k_D k_{\text{dimer}}[\text{L}]^2} \tag{4.44}$$

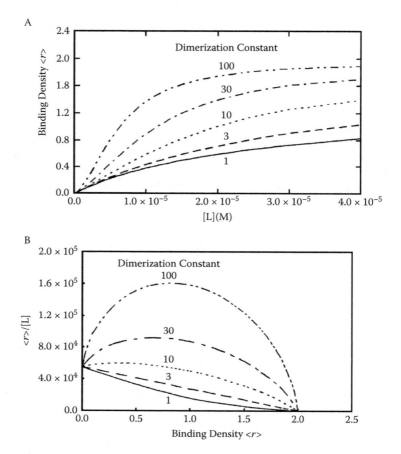

FIGURE 4.15 Cooperativity induced by ligand dimerization. Simulated binding isotherms for a macromolecule with two binding sites, with $k_1 = 5 \times 10^4$ M^{-1}, $k_2 = 5 \times 10^3$ M^{-1}, and selected values for k_{dimer} as shown. A. Direct plot. B. Scatchard plot.

When the ligand aggregates to an n-mer, and only the n-mer is bound to any appreciable extent, one then has

$$\mathbf{P} = 1 + K^{(n)}[L]^n, \quad <r> = \frac{nK^{(n)}[L]^n}{1 + K^{(n)}[L]^n} \tag{4.45}$$

This is essentially the model for all-or-none binding of n molecules of ligand to a receptor with n identical subunits, a model seen before, and which is well known to produce sigmoid binding curves.

4.5.2 AGGREGATION OF RECEPTOR AS A SOURCE OF COOPERATIVITY

A notable example of a ligand-binding system with an underlying receptor aggregational equilibrium is found with hemoglobin. The tetrameric $\alpha_2\beta_2$ complex, the form that is usually thought of as "hemoglobin," is actually in equilibrium with $\alpha\beta$ dimers. It requires a relatively concentrated solution of hemoglobin to drive the equilibrium toward the tetramer side so that the tetrameric species dominates. The dimer shows no detectable cooperativity in binding oxygen, unlike the tetramer, and the assembled tetramer has a reduced affinity for oxygen by comparison to the dimer (see Figure 4.16). Since the tetramer dominates at higher protein concentrations, the result is that, in a concentrated hemoglobin solution, the uptake of oxygen follows a sigmoid curve (due to positive cooperativity within the tetramer). Binding saturation in this concentrated protein solution requires, however, a higher concentration of oxygen than for a dilute protein solution that contains mostly the dimeric form. Thus the dimer-tetramer equilibrium leads to a dependence of hemoglobin oxygenation curves on the concentration of protein [45], and, even at millimolar concentrations of heme, the dimer-tetramer equilibrium must be taken into account for a precise model of the oxygenation process [46–48].

The following formal treatment of aggregation and binding cooperativity will trace the path laid out by Nichol et al. [43]; it will assume that the polymeric enzyme is composed of exactly n identical subunits in equilibrium with free monomer (that is, there is a single oligomerized species), and that there is random binding of ligand over the available sites on both monomer and polymer. This leads to a binding polynomial of the form

$$\mathbf{P} = (1 + k_1[\mathrm{L}])^p + K_{agg}[\mathrm{A}]^{n-1}(1 + k_2[\mathrm{L}])^q \tag{4.46}$$

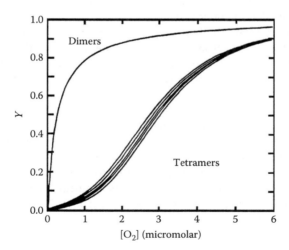

FIGURE 4.16 Effects on hemoglobin oxygenation induced by assembly of $\alpha\beta$ dimers into $\alpha_2\beta_2$ tetramers. Calculated isotherms for oxygenation at pH 8.5 as a function of [O_2] and protein concentration, for protein concentrations from 2.2 to 31 μM. Adapted from *Biochemistry* Vol. 23, A.H. Chu et al., "Effects of protons on the oxygenation-linked subunit assembly in human hemoglobin," pages 604–617, with permission. Copyright 1984 American Chemical Society.

and a binding equation:

$$< r > = \frac{pk_1[L](1+k_1[L])^{p-1} + qk_2[L]K_{agg}[A]^{n-1}(1+k_2[L])^{q-1}}{(1+k_1[L])^p + K_{agg}[A]^{n-1}(1+k_2[L])^q} \qquad (4.47)$$

Here [A] is the concentration of free monomeric receptor, K_{agg} is the equilibrium constant for forming the n-mer complex from n identical monomers, k_1 and k_2 are the site-binding constants for ligand on the monomer and on the n-mer aggregate, p is the number of binding sites per mole of monomer, and q is the number of binding sites per mole of n-mer complex. The model allows for exposure or occlusion of binding sites through aggregation of the monomeric units, depending on the relative values of np and q. A further extension of this model allows for multiple oligomeric species to be present in equilibrium with free monomeric receptor.

Notice that the concentration of free monomer [A] appears in the expressions for **P** and $< r >$, with [A] raised to the power $(n - 1)$. If n is set equal to unity, one easily recovers a simple isomerization model, though with possibly different numbers of sites in the two isomers. When n is 2 or greater, the dependence on free monomer concentration can produce binding cooperativity. Figure 4.17 shows the effects of receptor dimerization in both the direct plot format and the Scatchard format, for a fixed concentration of free monomeric receptor. The sigmoid character of the binding curve in the direct plot format is apparent, and the Scatchard plot shows the characteristic hump expected for positive binding cooperativity. Clearly, receptor aggregation can produce cooperative effects that strongly resemble those derived from allosteric models of cooperativity.

If the binding density (enzymatic catalytic rate, rate of uptake or release of drug or competitor, etc.) varies with receptor concentration, then receptor aggregation may play a role in regulating activity. However, simple thermodynamic nonideality should also be considered and eliminated before proposing changes in intrinsic binding activity that depend on aggregation state. An example of this can be found with hemoglobin, which at low protein concentrations has an activity approximately equal to the concentration. At the typical intracellular protein concentrations of around 340 g/L, however, molecular crowding raises the activity to about 50 times the hemoglobin concentration. This greatly affects the solubility of the protein, and is thought to be an important factor in the polymerization and gelation of hemoglobin S [49,50]; certainly, this nonideal behavior should be taken into account when considering the oxygenation curve for hemoglobin S under physiological conditions.

4.5.3 LIGAND DIMERIZATION DRIVEN BY A PIGGY-BACK LIGAND

Next, consider a combination of ligand aggregation and piggy-back binding into a single model for a cooperative binding effect. Here the receptor has two adjacent binding sites (a and b) that bind monomeric ligand (binding constants k_a and k_b) and that also may bind dimeric ligand (binding constant k_D). The dimer can bind a

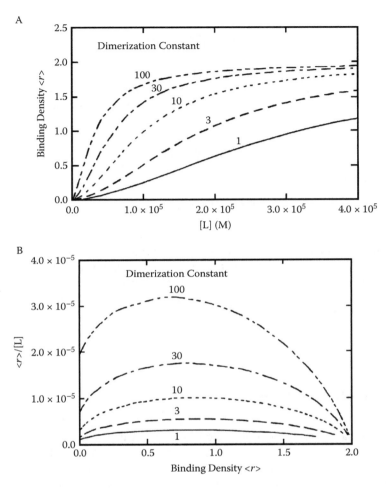

FIGURE 4.17 Cooperativity induced by receptor dimerization, with $k_1 = 1 \times 10^4$ M^{-1}, $k_2 = 1 \times 10^3$ M^{-1}, free monomeric receptor concentration [A] $= 1 \times 10^3$ M^{-1}, and values for the aggregation equilibrium constant K_{agg} as shown. A. Direct plot. B. Scatchard plot.

secondary, piggy-back ligand (denoted here as C), and then bind to the receptor with binding constant k_{DC}. The piggy-back ligand itself has no binding affinity for the receptor and it binds *only* to the dimer D, with a binding constant k_{p-back}. The binding polynomial is

$$\mathbf{P} = 1 + (k_a + k_b)[\mathrm{L}] + k_\mathrm{D}[\mathrm{D}] + k_{\mathrm{DC}}[\mathrm{DC}] \tag{4.48}$$

In terms of free monomer ligand and piggy-back ligand concentrations, this is

$$\mathbf{P} = 1 + (k_a + k_b)[\mathrm{L}] + k_{\mathrm{dimer}}[\mathrm{L}]^2[k_\mathrm{D} + k_{p-back}k_{\mathrm{DC}}[\mathrm{C}]] \tag{4.49}$$

or

$$\mathbf{P} = 1 + K^{(1)}[L] + K^{(2)}[L]^2 \qquad (4.50)$$

where the dependence on [C] of the phenomenological binding constant $K^{(2)}$ should be remembered. If the biological effect is due primarily to receptors with dimers bound, then the binding density $<r_2>$ of such states can be written as

$$<r_2> = \frac{K^{(2)}[L]^2}{1 + K^{(1)}[L] + K^{(2)}[L]^2} \qquad (4.51)$$

$K^{(2)}$ is given by

$$K^{(2)} = k_{\mathrm{dimer}}[k_D + k_{p-back}k_{DC}[C]] \qquad (4.52)$$

If [C] is zero, then this reduces to the previous case of simple ligand dimerization. The piggy-back ligand C can increase binding by increasing the concentration of dimers in solution. Furthermore, if it changes the conformation of the dimer such that the DC complex has a higher affinity than dimer D alone, then there is yet greater binding. This may, for example, be a mechanism by which the Tax protein of human T-cell leukemia virus increases the DNA binding of certain transcription factors that contain a basic region-leucine zipper (bZIP) DNA-binding domain [51].

4.6 NEGATIVE COOPERATIVITY

The examples of cooperativity presented so far have been mainly of the positive kind; there has as yet been no detailed discussion of negative cooperativity. However, negative cooperativity may play just as important a role in biological systems as does positive cooperativity. For a system with positive cooperativity, the sigmoid nature of the response curve allows for minimal response until a certain "threshold" concentration of ligand is reached, whereupon the system responds strongly, with a sharp rise in activity over a relatively narrow concentration range. This should be compared to the response for a system with independent sites, in which there is no "threshold" for response, and in which the rise in activity occurs over a broader concentration range. Negative cooperativity will dull this response yet further, spreading the response over an even greater concentration range and making it even more difficult to reach saturation. Positive cooperativity allows a large response to small changes in concentration; negative cooperativity tends to keep the activity relatively constant over substantially larger swings in concentration. That is, negative cooperativity can desensitize the system to fluctuations in concentration, whereas positive cooperativity can produce high sensitivity to small perturbations in ligand or substrate concentration.

For enzymes, with catalysis and rates of reactions being affected by ligand or substrate binding, Levitzki and Koshland [30] suggest some further reasons for the appearance of negative cooperativity:

1. Negative cooperativity is distinguished by a relatively high affinity at low degrees of site occupancy, and a decrease in affinity as the site occupancy rises. The high affinity of the first site or sites to be occupied keeps them filled with substrate (or ligand) most of the time, so that some reaction or response is possible even at low concentrations of substrate or ligand. A disadvantage of having only high affinity sites would likely be the slow dissociation kinetics that commonly accompany high binding affinity. Having available another population of sites with lower affinity (and faster dissociation kinetics) would allow turnover of substrate (or release of ligand) and so avoid a kinetic bottleneck due to slow dissociation.

2. Negative feedback is a common regulatory mechanism for metabolic pathways, in which the end product of a path is used as an allosteric effector to shut down a key enzyme near the beginning of the path. For an enzyme whose immediate product is important for several biochemical pathways, it may be advantageous to reduce flux through one pathway while maintaining at least a low degree of flux through another path. This can be accomplished by employing negative cooperativity in the binding of the allosteric effector. As the concentration of the effector ligand rises, overall pathway flux is indeed reduced, but thanks to the negative binding cooperativity, some activity will persist, and some product will be made for the other paths, even at very high effector concentrations.

3. For some enzymes, the free energy of ligand or substrate binding at one site can be used to drive a conformational change at a different site and so enhance reactivity there (presumably so long as the first site is occupied). The gain in catalytic efficiency at low substrate concentrations may be sufficient to offset the loss of activity due to occupation at the original site. The extreme here is a pattern called "half-of-the-sites reactivity," in which binding of substrate at one catalytic site is required in order to activate a second catalytic site. Levitzki and Koshland [30] present several examples of enzyme systems exhibiting this kind of behavior.

Negative cooperativity can arise from conformational changes, just as with positive cooperativity. It can also arise from aggregational phenomena and from piggyback binding of a second ligand species (see Section 4.5). Of course, just as with positive cooperativity there can be both homotropic and heterotropic ligand effects. Interestingly, it is possible to have "mixed" systems, in which the binding is positively cooperative over one range of ligand concentrations, and negatively cooperative over another range; Levitzki and Koshland [30] have tabulated some examples. Incidentally, apparent negative cooperativity can also derive from artifacts, such as preparations containing partially denatured receptor, or from carryover of tightly bound ligands through the preparation process. One might also find a sort of "accidental" negative cooperativity through the use of a very large "small" molecule effector that not only binds to one site but also occludes a nearby site as well. Finally, one obvious source of negative cooperativity arises when dealing with charged ligands, as with protons or other ions: binding of the first (charged) ligand may introduce electrostatic repulsions that hinder further addition of ligands.

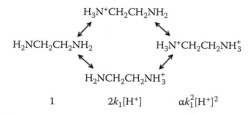

FIGURE 4.18 Protonation states of ethylene diamine, with their statistical weights.

4.6.1 NEGATIVE COOPERATIVITY IN THE TITRATION OF ETHYLENE DIAMINE

The titration of ethylene diamine offers a simple example of negative cooperativity due to electrostatic interactions (Figures 4.18 and 4.19). The titration proceeds from the neutral, un-ionized form through the singly charged form to the final doubly charged form. The introduction of the first positive charge onto the diamine reduces the affinity of the second amino group for a proton, hence the two inflection points in the titration curve. There is significant "resistance" to the addition of the second proton; that is, this system shows appreciable negative cooperativity.

The overall curve may be fit with a binding equation of the form

$$<r> = \frac{\kappa_1[H^+] + 2\kappa_1\kappa_2[H^+]^2}{1 + \kappa_1[H^+] + \kappa_1\kappa_2[H^+]^2} \tag{4.53}$$

with $\kappa_1 = 10^{9.93}$ M^{-1} and $\kappa_2 = 10^{6.85}$ M^{-1} [52]; these two parameters are reciprocals of ionization constants determined by fitting to experimental titration curves. In terms of individual site binding constants the binding equation could be written as

$$<r> = \frac{2k_1[H^+] + 2\alpha k_1^2[H^+]^2}{1 + 2k_1[H^+] + \alpha k_1^2[H^+]^2} \tag{4.54}$$

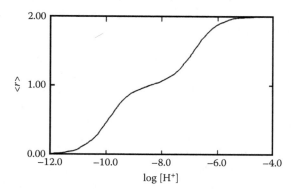

FIGURE 4.19 Titration curve for ethylene diamine, using stepwise binding constants $\kappa_1 = 10^{9.93}$ M^{-1} and $\kappa_2 = 10^{6.85}$ M^{-1} [52].

and with some algebra one finds that $k_1 = \frac{1}{2} \kappa_1 = 10^{9.63}$ M^{-1} and $\alpha = 2 \kappa_2/k_1 = 10^{-2.48}$ $= 3.3 \times 10^{-3}$. This value for the (negative) cooperativity corresponds to an unfavorable free energy change of about 14 kJ/mol at room temperature, quite a significant hindrance to the addition of a second ligand.

4.6.2 GLYCERALDEHYDE 3-PHOSPHATE DEHYDROGENASE AND NEGATIVE COOPERATIVITY

The enzyme glyceraldehyde-3-phosphate dehydrogenase (GPDH) from rabbit muscle is a well-studied enzyme that provides an interesting example of negative cooperativity. This tetrameric enzyme catalyzes the conversion of glyceraldehyde 3-phosphate to 1,3-bisphosphoglycerate, with the concomitant reduction of the nicotinamide cofactor NAD$^+$ to NADH. A variety of experimental techniques (equilibrium dialysis, ultraviolet absorbance, and fluorimetric assays) have shown a reduction in catalytic activity and nicotinamide binding affinity with increasing concentrations of NAD$^+$ or of analogues of the nicotinamide cofactor [53–57]. The stoichiometry of binding (four nicotinamides per enzyme) is consistent with the composition of the enzyme as a tetramer of four identical subunits. The observed reduction in binding affinity for NAD$^+$ could be interpreted as the result of negative cooperativity. However, the enzyme might instead simply have two (or more) preexisting classes of binding sites that acted independently and that simply differed in their affinity toward NAD$^+$ [58,59]. Crystallographic data for the rabbit muscle enzyme originally could not resolve the question of different preexisting classes of sites versus identical sites that interacted with negative cooperativity. In an elegant application of binding thermodynamics, Henis and Levitzki [60,61] showed how binding competition experiments could be used to distinguish between the two competing models.

Their plan of attack was to study the binding cooperativity of a primary ligand species and how it depended on the binding of a competing ligand. The cooperativity was to be characterized either by changes in the Hill parameter, or through detailed isotherm curve fitting and analysis of the resulting phenomenological site binding constants, according to the Adair equation. It was found that ATP, adenosine diphosphoribose (ADP-Rib), and acetylpyridine adenine dinucleotide (AcPyAD$^+$) bound to the enzyme independently and without any cooperativity. On the other hand, NAD$^+$ and its analog nicotinamide-1-N^6-ethenoadenine dinucleotide (εNAD$^+$) bound with apparent negative cooperativity, or to two classes of two independent sites each (Figure 4.20). However, because of the very high affinity of the enzyme for the first two equivalents of NAD$^+$, it was not possible to obtain the full binding curve for the NAD$^+$ ligand alone, and instead the investigation supplemented this study with work on the binding of the etheno analog εNAD$^+$, for which full and detailed isotherms could be readily obtained.

The reasoning behind the competition assay was as follows. Suppose that the enzyme indeed has two classes of two noninteracting sites each, with respect to a ligand X. The Hill parameter at half saturation $n_H(X)$ for binding of this ligand species is

$$n_H(X) = \frac{4}{2 + (K_X' / K_X'')^{1/2} + (K_X'' / K_X')^{1/2}} \qquad (4.55)$$

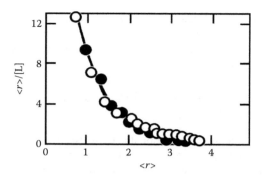

FIGURE 4.20 Negative cooperativity in the binding of ε-NAD⁺ to glyceraldehyde-3-phosphate dehydrogenase from rabbit muscle. ● and ○ represent two separate experiments. Line represents suggested fit to selected model. Adapted from *Proceedings of the National Academy of Sciences USA* Vol. 77, Y.I. Henis and A. Levitzki, "Mechanism of negative cooperativity in glyceradehyde-3-phosphate dehydrogenase deduced from ligand competition experiments," with permission. Copyright 1980.

where K'_X and K''_X are binding constants for X to the two classes of sites. Now suppose that a competing ligand species Z is added, at concentrations much higher than the concentration of binding sites (this will ensure an essentially constant concentration of Z during the titration). The Hill parameter at 50% saturation with respect to X in the presence of Z is now

$$n_H(X, Z) = \cfrac{4}{2 + \cfrac{K'_X + K''_X + (K'_X K''_Z + K'_Z K''_X)[Z]}{(K'_X K''_X)^{1/2}(1 + (K'_Z + K''_Z)[Z] + K'_Z K''_Z[Z]^2)^{1/2}}} \tag{4.56}$$

where K'_Z and K''_Z denote the two binding constants for Z to the two classes of sites. If Z binds without any cooperativity, then $K'_Z = K''_Z$ and the formula for $n_H(X,Z)$ reduces simply to the previous one for $n_H(X)$. This gives the notable prediction that $n_H(X, Z) = n_H(X)$; that is, the Hill parameter remains unchanged.

Now suppose that in binding ligand X the enzyme follows a simple KNF model with its associated cooperativity. The simplest case here would be a tetrameric enzyme composed of two noninteracting dimers, but with ligand-induced conformational changes in each dimer. (More elaborate models could of course be constructed, but their analysis would lead to the same qualitative predictions as this simple model of negative cooperativity.) The unliganded subunit conformation will be denoted as A, and the X-liganded conformation as B. The Hill parameter for this model is

$$n_H(X) = \frac{2}{1 + \left(K_{AB}^2 / K_{BB}\right)^{1/2}} \tag{4.57}$$

where K_{AB} and K_{BB} are the subunit interaction equilibrium constants. Now suppose ligand Z is added, again at sufficiently high concentrations that ensure an essentially constant [Z]; for further generality the model will allow for a third possible conformation of the subunits, denoted as C; and it allows for conversion from conformation A or B to C. Now titrate with species X. The Hill parameter in this case is now

$$
n_H(X, Z) = \cfrac{2}{1 + \left[\cfrac{K_{AB}^2 (1 + K_{BC} K_{Z\text{-}C} K_{t\text{-}AC}[Z] / K_{AB})^2}{K_{BB}(1 + 2K_{AC} K_{Z\text{-}C} K_{t\text{-}AC}[Z] + K_{CC} \left(K_{Z\text{-}C} K_{t\text{-}AC} \right)^2 [Z]^2)} \right]^{1/2}}
\qquad (4.58)
$$

where K_{AC} and K_{BC} describe the interaction between an unliganded subunit (A) and a Z-liganded subunit (C), and between an X-liganded subunit (B) and a Z-liganded subunit, while K_{CC} applies to the interaction between two Z-liganded subunits. $K_{Z\text{-}C}$ describes the affinity of Z for a subunit in the C conformation, and $K_{t\text{-}AC}$ describes the transformation of a subunit from conformation A to C.

If Z binds without cooperativity, then this model requires $(K_{AC}^2/K_{CC}) = 1$. Notice, however, that the ratio (K_{BC}/K_{CC}) can take on any value, and so $n_H(X,Z)$ is not necessarily equal to $n_H(X)$, even though Z is binding without cooperativity. This is quite a distinct prediction from that provided by the model of two classes of independent sites; there the two values of the Hill parameter (for the absence and presence of Z) had to match if Z bound without cooperativity, but here the two values are allowed to differ. If experimental data were to show that the Hill parameter indeed changed upon introduction of the ligand Z, this would be a strong argument against the model of preexisting independent classes of sites. It would not "prove" the correctness of the particular version of the sequential model under discussion here, but it would definitely be evidence in favor of a model with negative cooperativity.

Henis and Levitzki proceeded to evaluate the Hill parameter for the enzyme binding NAD^+ and εNAD^+ in the absence and presence of a variety of competing ligands. Their results are presented in Table 4.1 [60]. Note that the Hill coefficient for binding of εNAD^+ alone is 0.64, consistent with either negative cooperativity or multiple classes of independent sites. As mentioned before, the investigators were unable to fully evaluate the isotherm to obtain a Hill coefficient for NAD^+ itself, but they were able to calculate the Hill coefficient in the presence of certain of the competing ligands.

TABLE 4.1
Hill Parameters for Glyceraldehyde-3-Phosphate Dehydrogenase Binding NAD^+ and εNAD^+

Competing Ligand	n_H for εNAD^+	n_H for NAD^+
None	0.64 ± 0.04	N/A
ADP-ribose	0.65 ± 0.04	0.58 ± 0.03
AcPyAD$^+$	1.00 ± 0.04	N/A
ATP	0.89 ± 0.02	0.88 ± 0.04

From Henis and Levitzki [60].

They also showed that ATP and AcPyAD$^+$ bound without any cooperativity to the enzyme.

Using the competing ligands ATP and AcPyAD$^+$ with εNAD$^+$, Henis and Levitzki found that both ATP and AcPyAD$^+$ significantly reduced the apparent (negative) cooperativity found with εNAD$^+$ alone (that is, the Hill parameter in the presence of these ligands was greater than in their absence). This strongly implied that the enzyme did indeed bind εNAD$^+$ with negative cooperativity. (The model of multiple classes of independent sites on the enzyme cannot explain the observed change in the Hill parameter for εNAD$^+$ in the presence of these ligands.) Interestingly, in the presence of the competing ligands ATP and ADP-ribose (a ligand that also binds without cooperativity), the values of the Hill parameter for NAD$^+$ were quite distinct, indicating that they have different effects on the cooperativity of binding of NAD$^+$. Also, the Hill parameter values for εNAD$^+$ and NAD$^+$ matched relatively closely, supporting the idea that NAD$^+$ itself binds with negative cooperativity.

Taken all together, these results showed that the rabbit-derived enzyme was binding its nicotinamide cofactors with strongly negative cooperativity. The data clearly did not support the model of previously existing multiple classes of sites; and the alternative proposal of ligand-induced, sequential changes in subunit conformation, with negative binding cooperativity, was strongly supported. This is a good example of the insightful application of binding thermodynamics to distinguish between alternative binding mechanisms.

REFERENCES

1. Bohr, C., Hasselbalch, K.A., and Krogh, A., Ueber einen in biologischer Beziehung wichtigen Einfluss, den die Kohlensäurespannung des Blutes auf dessen Sauerstoffbindung übt, *Skand. Arch. Physiol.* 16, 402, 1904.
2. Svedberg, T. and Fåhreus, R., A new method for the determination of the molecular weight of the proteins, *J. Am. Chem. Soc.* 48, 430, 1926.
3. Adair, G.S., A critical study of the direct method of measuring the osmotic pressure of hemoglobin, *Proc. Roy. Soc. London A* 108A, 627, 1925.
4. Adair, G.S., The hemoglobin system. VI. The oxygen dissociation curve of hemoglobin, *J. Biol. Chem.* 63, 529, 1925.
5. Pauling, L., The oxygen equilibrium of hemoglobin and its structural interpretation, *Proc. Natl. Acad. Sci. USA* 21, 186, 1935.
6. Perutz, M.F. et al., Structure of haemoglobin. A three-dimensional Fourier synthesis at 5.5-Å resolution, obtained by X-ray analysis, *Nature* 185, 416, 1960.
7. Perutz, M.F., Structure and mechanism of haemoglobin, *Br. Med. Bull.* 32, 195, 1976.
8. Perutz, M.F., Regulation of oxygen affinity of hemoglobin: Influence of structure of the globin on the heme iron, *Annu. Rev. Biochem.* 48, 327, 1979.
9. Gerhart, J.C. and Pardee, A.B., The enzymology of control by feedback inhibition, *J. Biol. Chem.* 237, 891, 1962.
10. Kantrowitz, E.R. and Lipscomb, W.N., *Escherichi coli* aspartate transcarbamylase: The relation between structure and function, *Science* 241, 669, 1988.
11. Lipscomb, W.N., Aspartate transcarbamylase from *Escherichia coli*: Activity and regulation, *Adv. Enzymol.* 68, 67, 1994.
12. Monod, J., Wyman, J., and Changeux, J.-P., On the nature of allosteric transitions: A plausible model, *J. Mol. Biol.* 12, 88, 1965.

13. Koshland, D.E. Jr., Nemethy, G., and Filmer, D., Comparison of experimental binding data and theoretical models in proteins containing subunits, *Biochemistry* 5, 365, 1966.
14. Cornish-Bowden, A., *Fundamentals of Enzyme Kinetics*. London: Portland Press, 1995, p. 234.
15. Monod, J. and Jacob, F., General conclusions: Teleonomic mechanisms in cellular metabolism, growth, and differentiation, *Cold Spring Harbor Symp. Quant. Biol.* 26, 389, 1961.
16. Monod, J., Changeux, J.-P., and Jacob, F., Allosteric proteins and cellular control systems, *J. Mol. Biol.* 6, 306, 1963.
17. Wyman, J., Allosteric effects in hemoglobin, *Cold Spring Harbor Symp. Quant. Biol.* 28, 483, 1963.
18. Hill, A.V., The possible effects of the aggregation of the molecules of haemoglobin on its dissociation curves, *J. Physiol.* 40, iv, 1910.
19. Wyman, J., Linked functions and reciprocal effects in hemoglobin: A second look, *Adv. Protein Chem.* 19, 223, 1964.
20. Cornish-Bowden, A. and Koshland, D.E. Jr., Diagnostic uses of the Hill (logit and Nernst) plots, *J. Mol. Biol.* 95, 201, 1975.
21. Schellman, J.A., Fluctuation and linkage relations in macromolecular solution, *Biopolymers* 29, 215, 1990.
22. Holt, J.M. and Ackers, G.K., Asymmetric distribution of cooperativity in the binding cascade of normal human hemoglobin. 2. Stepwise cooperative free energy, *Biochemistry* 44, 11939, 2005.
23. Watari, H. and Isogai, Y., A new plot for allosteric phenomena, *Biochem. Biophys. Res. Commun.* 69, 15, 1976.
24. Whitehead, E.P., Co-operativity and the methods of plotting binding and steady-state kinetic data, *Biochem. J.* 171, 501, 1978.
25. Klotz, I.M., Ligand-Receptor Energetics: A Guide for the Perplexed. New York: Wiley, 1997, p. 49.
26. Cook, R.A. and Koshland, D.E. Jr., Positive and negative cooperativity in yeast glyceraldehyde 3-phosphate dehydrogenase, *Biochemistry* 9, 3337, 1970.
27. Haurowitz, F., Das Gleichgewicht zwischen Hämoglobin und Sauerstoff, *Hoppe-Seyler's Z. Physiol. Chem.* 254, 266, 1938.
28. Koshland, D.E., Application of a theory of enzyme specificity to protein synthesis, *Proc. Natl. Acad. Sci. USA* 44, 98, 1958.
29. Hammes, G.G. and Wu, C.-W., Regulation of enzyme activity, *Science* 172, 1205, 1971.
30. Levitzki, A. and Koshland, D.E., The role of negative cooperativity and half-of-the-sites reactivity in enzyme regulation, *Curr. Topics Cell Reg.* 10, 1, 1976.
31. Gill, S.J. et al., Oxygen binding constants for human hemoglobin tetramers, *Biochemistry* 26, 3995, 1987.
32. Di Cera, E., Robert, C.H., and Gill, S.J., Allosteric interpretation of the oxygen-binding reaction of human hemoglobin tetramers, *Biochemistry* 26, 4003, 1987.
33. Lukin, J.A. and Ho, C., The structure-function relationship of hemoglobin in solution at atomic resolution, *Chem. Rev.* 104, 1219, 2004.
34. Ackers, G.K. and Holt, J.M., Asymmetric cooperativity in a symmetric tetramer: Human hemoglobin, *J. Biol. Chem.* 281, 11441, 2006.
35. Robert, C.H. et al., Nesting: Hierarchies of allosteric interactions, *Proc. Natl. Acad. Sci. USA* 84, 1891, 1987.
36. Decker, H. et al., Nested allosteric interaction in tarantula hemocyanin revealed through the binding of oxygen and carbon monoxide, *Biochemistry* 27, 6901, 1988.

37. Yifrach, O. and Horovitz, A., Nested cooperativity in the ATPase activity of the oligomeric chaperonin GroEL, *Biochemistry* 34, 5302, 1995.

38. Traut, T.W., Dissociation of enzyme oligomers: A mechanism for allosteric regulation, *CRC Crit. Rev. Biochem. Mol. Biol.* 29, 125, 1994.

39. Metzger, H. and Kinet, J.-P., How antibodies work: Focus on Fc receptors, *FASEB J.* 2, 3, 1988.

40. Pawson, T., Protein modules and signalling networks, *Nature* 373, 573, 1995.

41. Pawson, T. and Scott, J.D., Signaling through scaffold, anchoring, and adaptor proteins, *Science* 278, 2075, 1997.

42. Frieden, C., Treatment of enzyme kinetic data. II. The multisite case: Comparison of allosteric models and a possible new mechanism, *J. Biol. Chem.* 242, 4045, 1967.

43. Nichol, L.W., Jackson, W.J.H., and Winzor, D.J., A theoretical study of the binding of small molecules to a polymerizing protein system. A model for allosteric effects, *Biochemistry* 6, 2449, 1967.

44. Kurganov, B.I., Kinetic analysis of dissociating enzymatic systems (English translation), *Molek. Biol.* 2, 351, 1968.

45. Mills, F.C., Johnson, M.L., and Ackers, G.K., Oxygenation-linked subunit interactions in human hemoglobin: Experimental studies on the concentration dependence of oxygenation curves, *Biochemistry* 15, 5350, 1976.

46. Johnson, M.L., Halvorson, H.R., and Ackers, G.K., Oxygenation-linked subunit interactions in human hemoglobin: Analysis of linkage functions for constituent energy terms, *Biochemistry* 15, 5363, 1976.

47. Mills, F.C. and Ackers, G.K., Quaternary enhancement in binding of oxygen by human hemoglobin, *Proc. Natl. Acad. Sci. USA* 76, 273, 1979.

48. Chu, A.H., Turner, B.W., and Ackers, G.K., Effects of protons on the oxygenation-linked subunit assembly in human hemoglobin, *Biochemistry* 23, 604, 1984.

49. Ross, P.D. and Minton, A.P., Analysis of non-ideal behavior in concentrated hemoglobin solutions, *J. Mol. Biol.* 112, 437, 1977.

50. Ross, P.D. and Minton, A.P., The effect of non-aggregating proteins upon the gelation of sickle cell hemoglobin: Model calculations and data analysis, *Biochem. Biophys. Res. Commun.* 88, 1308, 1979.

51. Baranger, A.M. et al., Mechanism of DNA-binding enhancement by the human T-cell leukaemia virus transactivator Tax, *Nature* 376, 606, 1995.

52. Butler, J.N., *Ionic Equilibrium: A Mathematical Approach.* Reading, MA: Addison-Wesley, 1964.

53. Conway, A. and Koshland, D.E., Negative cooperativity in enzyme action. The binding of diphosphopyridine nucleotide to glyceraldehyde 3-phosphate dehydrogenase, *Biochemistry* 7, 4011, 1968.

54. Devijlder, J.J.M. and Slater, E.C., The reaction between NAD^+ and rabbit-muscle glyceraldehydephosphate dehydrogenase, *Biochim. Biophys. Acta* 167, 23, 1968.

55. Schlessinger, J. and Levitzki, A., Molecular basis of negative co-operativity in rabbit muscle glyceraldehyde-3-phosphate dehydrogenase, *J. Mol. Biol.* 82, 547, 1974.

56. Bell, J.E. and Dalziel, K., Studies of coenzyme binding to rabbit muscle glyceraldehyde-3-phosphate dehydrogenase, *Biophys. Biochim. Acta* 391, 249, 1975.

57. Scheek, R.M. and Slater, E.C., Preparation and properties of rabbit-muscle glyceraldehyde-phosphate dehydrogenase with equal binding parameters for the third and fourth NAD^+ molecules, *Biochim. Biophys. Acta* 526, 13, 1978.

58. Malhotra, O.P. and Bernhard, S.A., Spectrophotometric identification of an active site-specific *acyl* glyceraldehyde 3-phosphate dehydrogenase, *J. Biol. Chem.* 243, 1243, 1968.
59. MacQuarrie, R.A. and Bernhard, S.A., Mechanism of alkylation of rabbit muscle glyceraldehyde 3-phosphate dehydrogenase, *Biochemistry* 10, 2456, 1971.
60. Henis, Y.I. and Levitzki, A., The sequential nature of the negative cooperativity in rabbit muscle glyceraldehyde-3-phosphate dehydrogenase, *Eur. J. Biochem.* 112, 59, 1980.
61. Henis, Y.I. and Levitzki, A., Mechanism of negative cooperativity in glyceraldehyde-3-phosphate dehydrogenase deduced from ligand competition experiments, *Proc. Natl. Acad. Sci. USA* 77, 5055, 1980.

5 Binding to Lattices of Sites

5.1 LINEAR LATTICES OF BINDING SITES

A number of important biopolymer and synthetic polymer systems form what are essentially one-dimensional arrays of binding sites. The polymer chain can be idealized as a linear array or lattice of monomeric units that are potential binding sites for ligands such as drugs, protons, and proteins. A familiar example of such a linear lattice would be a polynucleotide, whose bases or base pairs serve as the points of contact to bind polyamines, dye molecules, proteins, etc. Many other linear biopolymers (proteins, polysaccharides, etc.) of course might be idealized as linear lattices of sites, too. The exact definition of a *site*, however, will depend on the nature of the ligand as well as on the composition and sequence of the monomeric units in the polymer. For example, with double-stranded polynucleotides it may be expedient in some cases to think of the phosphates of the phosphodiester backbone as forming the lattice of sites, especially when the ligand-polynucleotide interaction is primarily through ionic attraction. In other cases, in which hydrogen bonding to particular base sequences is involved, for example, it may be better to view as the binding sites the bases or base pair units that offer such points for interaction. And yet again, when considering the intercalation of planar aromatic ligands, such as ethidium, into double-stranded DNA, one can think of the site as being formed by adjacent base pairs, as they unstack and unwind the helix to form a pocket for the planar ligand. An interesting complication here is the potential for overlap between binding sites for ligands that cover or interact with two or more adjacent monomeric units in the macromolecule.

Positive cooperative interactions among the bound ligands can become quite important in these linear lattice systems. A surprisingly weak degree of positive cooperativity can lead to nearly all-or-none coverage of the entire length of the polymer by the ligand. With heteromeric lattices (those with more than one type of site in the lattice), cooperativity can also amplify slight differences in sequence binding preferences. In some systems, conformational changes in the macromolecule can be driven by cooperative ligand binding, so that small changes in the ligand concentration result in gross conformational shifts in the polymer (e.g., a helix-to-coil transition).

In 1974 there appeared an important paper on the theory of binding of ligands to a one-dimensional lattice of uniform sites [1] that presented and analyzed an intuitive model for DNA-protein interactions. The McGhee-von Hippel paper was quite influential since it appeared in a leading journal just as a large number of biophysical chemists were starting to study DNA-protein interactions. Furthermore, the results were given in a format that could be immediately used in interpreting binding data. McGhee and von Hippel used combinatorics and conditional probabilities to derive their results. Their arguments are quite involved, but since the fundamental results may be derived by other, independent routes, there is no doubt about the theoretical validity of the conclusions. McGhee and von Hippel were primarily interested in describing DNA-protein interactions, but their model is general and can be applied to many other macromolecular binding systems that have the appropriate geometry. Because of this model's wide applicability, and because many other models build on it, it is worth taking the time to go into its details. Fortunately, the overall specifications and concepts behind the model are relatively clear and intuitive.

5.1.1 REDEFINING THE BINDING DENSITY FOR LINEAR SYSTEMS

Before presenting the details of the McGhee-von Hippel model, it is necessary to clarify some subtle points concerning the definitions of binding density and binding stoichiometry. When a ligand covers or makes contact with two or more sites on a linear lattice, there is the potential for overlap among the binding sites. If the sites overlap, how is the number of (potential) binding sites to be counted? Since the number of binding sites per macromolecule is needed to relate the binding density $<r>$ to the degree of binding saturation Y, it is also necessary to consider how to define the binding density appropriately for these linear lattice systems.

Consider a macromolecule with n binding sites, possibly independent (or not) and possibly of equal affinity (or not). The binding reaction at any stage of the titration can be written according to the law of mass action as

$$\text{Free ligand} + \text{Unoccupied site} \rightleftharpoons \text{Bound ligand} \qquad (5.1)$$

An apparent equilibrium binding constant Q could be defined here as a ratio of the bound ligand to the free ligand, scaled by the available binding sites:

$$Q = \frac{[\text{Bound ligand}]}{[\text{Free ligand}][\text{Unoccupied sites}]} \qquad (5.2)$$

To obtain an equation suitable for Scatchard-type analysis of the binding data, this last relation can be divided on both sides by the concentration of macromolecule and rearranged to get

$$\frac{<r>}{[L]} = Q \times F(n, <r>) \qquad (5.3)$$

where the binding density $<r>$ equals the concentration of bound ligand divided by the concentration of macromolecule, and [L] is the free ligand concentration. The function $F(n, <r>)$ accounts for the interconnection of binding density and number of sites. In the simple model with n equal and independent sites, the function $F(n, <r>)$ is equal to $(n - <r>)$ and Q is equal to the site binding constant K, which gives the familiar form

$$\frac{<r>}{[L]} = K(n - <r>) \qquad (5.4)$$

When there is cooperativity among the sites or there are multiple classes of binding sites, the functions Q and F are of course more complicated and will depend on the cooperativity, the different binding constants, etc.

The sites on a latticelike macromolecule like DNA or a polysaccharide like heparin are not physically separated from one another and so they are not necessarily independent. *Large ligands* will be defined as those that can make multiple contacts with such lattices and so occupy multiple lattice sites that are adjacent to one another. These large ligands may also exert a sort of site exclusion effect as they bind; their physical presence on the lattice may leave some adjacent sites vacant, but the vacancy is not large enough to accommodate another ligand. Such vacant sites will be referred to as *occluded sites*. The number of such vacant, occluded sites will vary throughout the titration as the ligands pack together on the lattice. This feature of site occlusion produces conceptual difficulties, however. Should these occluded sites be counted in the number of free binding sites and so appear in the denominator in the definition of the effective binding constant Q? Or are these sites occupied, at least in the sense that another ligand cannot directly occupy them, so that their number should be included in the numerator for Q? And how does this affect the overall number n of binding sites on the macromolecule?

Consider a dimeric ligand binding to a linear lattice of N sites, and suppose that, for simplicity, N is even. In binding the first dimer there are $(N - 1)$ different places along the lattice where it may be placed, without letting the ligand leave one end or the other "dangling" off the lattice. This suggests that the number of binding sites in this system is $(N - 1)$. But now consider the number of ligands that can be placed on the lattice at saturation—this is $N/2$, which suggests that the number of binding sites must be $N/2$, which is certainly not the same as $(N - 1)$. Which is the true number of sites? A sketch of the situation for a simplified model may help to clarify these questions; see Figure 5.1, where N equals 8.

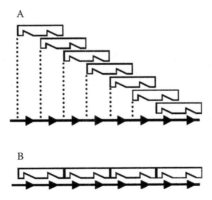

FIGURE 5.1 The problem of defining the binding stoichiometry for linear lattice systems. A. A single dimer can be placed $(N - 1)$ ways on an N-site lattice. Here, a dimer can be placed seven ways on a lattice with eight sites. B. A saturated eight-site lattice, with $N/2 =$ four dimers bound.

But now the definition of the binding density $< r >$ becomes less clear. A reasonable working definition of the degree of binding saturation, Y, is

$$Y = \frac{[\text{Bound ligand}]}{[\text{Bound ligand}]_{max}} \qquad (5.5)$$

where $[\text{Bound ligand}]_{max}$ is the amount bound at saturation of the macromolecule. Y is defined clearly enough in terms of the maximum number of ligands that can be bound by the macromolecule, so the question of vacant sites does not matter here. For systems in which the binding sites are physically separated, Y is equal to $< r >$ /n, or equivalently, $< r >$ equals nY. But if the number of binding sites n on a lattice is uncertain, what then of the binding density $< r >$?

It becomes necessary now to define a new quantity, the average number of ligands bound *per lattice site*, and give it the symbol $< v >$:

$$< v > = \frac{[\text{Bound ligand}]}{[\text{Total lattice sites}]} \qquad (5.6)$$

The quantity $< v >$ will be called the *site binding density*; the brackets serve as a reminder that this is an average quantity like the binding density itself. Experimentally, the exact lengths of the polymer chains in a sample are often not known, but it may be possible to find the concentration of, e.g., the DNA bases for a polynucleotide sample (or the concentration of acid groups for polyacrylic acid titrations, or the total glucose concentration for glycogen binding enzymes, etc.), so that there is a measure of the total number of lattice sites. Division of the amount of bound ligand by the total concentration of lattice sites keeps the quantity $< v >$ dimensionless, and allows comparison across different experiments.

The site binding density $< v >$ will always be less than or equal to 1. For ligands that occupy h contiguous sites on a lattice, $< v >$ will have a maximum value of $1/h$. The quantity h will be called the *site size* of the ligand. The relation between $< v >$ and the degree of binding saturation is then $Y = h < v >$.

It is easy to confuse the binding density $< r >$ with the site binding density $< v >$ for lattice systems. The quantity $< v >$ is in fact sometimes loosely referred to as the *binding density*, but $< v >$ has a different physical definition and a different mathematical range than that of the binding density $< r >$ itself. The site binding density concept applies specifically to binding to lattices, while $< r >$ is a more general measure of binding.

In discussing lattice binding it is necessary to take care to distinguish the number of lattice sites occupied by a single ligand (symbol h) from n, the number of binding sites as defined for a Scatchard-type analysis. There is the chance of notational confusion here, since McGhee and von Hippel used the symbol n for the number of contiguous lattice sites occupied by a ligand, and this usage has persisted in the literature. Here the symbol h for the site size parameter has been deliberately selected so as to avoid possible confusion of the two quantities.

5.1.2   THE MCGHEE-VON HIPPEL TREATMENT [1]

McGhee and von Hippel ignored chemical differences among binding sites (DNA base sequences, sequences of amino acids in a protein, etc.), and simplified the polymer (of whatever type) to a string of regularly spaced, identical units. A polymer of N monomer units is now represented by a linear or one-dimensional lattice of N sites. The sites of the lattice are identical, and will provide the places where a ligand might bind. McGhee and von Hippel assumed that the lattice length N was a large number; that is, only very long polymer chains were considered in their treatment. This was done to avoid having to deal with the end units as special cases.

When a drug or protein binds to DNA, it can physically occupy a volume (or length of polymer) extending over several base pairs. Contacts may be made with several adjacent bases or base pairs by a single bound ligand. In other words, a binding site may not be just a single base or base pair, but rather h adjacent polymer units. In terms of a lattice model of the polymer, the ligand occupies h of the lattice sites. Its occupation of these sites blocks or shields them from making contact with a second ligand; a second ligand may not make contact with the occupied sites.

The occupation of two or more adjacent sites by a single ligand can block not just those occupied sites, but it may also reduce the ways in which another ligand may bind to the vacant neighboring lattice sites. For example, the gap of vacant sites between two neighboring ligands may be too small to permit physical occupancy by a third ligand. The lattice now has fewer available *binding sites*, though of course the number of unoccupied lattice units is the same. The overall capacity of the lattice to accept another ligand has been reduced. This effect is referred to as *site exclusion*. It is basically an entropic effect, in which partial occupation of the lattice reduces the number of available sites beyond that expected for independent (nonlattice) site binding. The net result is a sort of negative cooperativity, effectively a repulsion between the ligands that reduces the overall lattice occupancy. If the ligand occupies only one lattice site, however, there is no site exclusion and a single "open" site on the lattice then fully qualifies as a potential binding site. This may be the case for very simple ligands, such as protons titrating a polyacid.

The binding of a ligand onto the lattice so that it is isolated (without immediate neighbors to either left or right) will define the basic equilibrium constant, K, to be used in the model. This binding constant is not assigned to any particular residue or lattice site in the h residues occupied by the ligand. Partial binding contacts, such that some number less than h sites are occupied, are disallowed in the model.

What about attractive or repulsive interactions between the bound ligands? Recall that a type of negative cooperativity can be induced if there is steric exclusion along the lattice by ligands that occupy two or more adjacent sites. But the model should also include the possibility of positive cooperativity, in which the binding of a first ligand makes more favorable the binding of adjacent ligands. The basic McGhee-von Hippel model limits itself to interactions between immediately adjacent ligand neighbors on the polymer. This limitation to nearest neighbors is done mainly to keep the mathematics simple. A treatment of binding effects from second-, third-, etc. nearest neighbors would introduce more parameters than might reasonably be justified from experimental binding data, in most cases. Exceptions might be made for ligands that are especially simple (e.g., a proton) or where the forces behind the cooperativity are long-range, as with the Coulomb force.

The model accounts for cooperativity by multiplying the binding constant of a second ligand, and binding next to the first bound ligand, by a suitable positive constant, ω. The quantity ω is referred to as the cooperativity parameter, or simply the cooperativity. This factor in the binding constant might be written as K_{coop}, which would clearly mark its role in binding, but (following McGhee and von Hippel) this book will use the symbol ω instead.

One useful way to view ω is that it is the equilibrium constant for moving an *isolated* bound ligand along the lattice to a place where it is *next to* another bound ligand. The extra free energy enjoyed by the transposed ligand is $\Delta G_{coop} = -RT \ln \omega$. Thus, one can think of ω as a multiplier of the intrinsic binding constant K that increases K to account for the favorable interaction with the neighbor. In other words, the binding constant for this second ligand occupying a site next to the first ligand is not just K but $K\omega$.

Since a bound ligand might have two nearest neighbors, one on either side, it is fair to ask what the free energy change might be for binding a ligand so that it interacts with both neighbors, simultaneously. Consider a polymer lattice with two bound ligands that are spaced such that there is a vacancy of exactly h lattice sites between them. Now bring a third ligand up to the polymer and bind it in exactly this site, so that it is flanked right and left by ligands, with no intervening vacancies on the lattice. For this configuration the change in free energy has three sources. First, there is the contribution from simple binding, equal to $-RT \ln K$ [L], where [L] is the concentration of free ligand in solution. Next, there is the cooperative interaction with the ligand on the left: $\Delta G_{coop} = -RT \ln \omega$. And finally, there is the cooperative interaction with the ligand on the right, with another contribution $\Delta G_{coop} = -RT \ln \omega$. Since the cooperative interaction with the left-side ligand multiplies K by the cooperativity ω, one ought to also multiply K by another factor of ω for the right-side ligand. So overall, the effective binding constant for the ligand in this situation is $K\omega^2$, not just $K\omega$. Figure 5.2 summarizes these conventions.

A value of 1.0 for ω means that there is just simple site exclusion operating among the ligands. This corresponds to the absence of any positive cooperativity, much as with statistical binding of ligands to identical, noninteracting sites on a multimeric protein, for example. This book will follow common practice of workers in this area and refer to the case in which $\omega = 1.0$ as constituting "noncooperative" binding. However, site exclusion on a linear lattice is itself inherently anticooperative in its effects; so that even with a value of 1.0 for ω, if a ligand occupies more than a single lattice site, then there will be site exclusion and a binding curve showing anticooperativity.

A value for ω greater than 1 implies a tendency for ligands to aggregate, to bind next to neighbors. Conversely, a value of ω less than 1 indicates an overt negative cooperativity (beyond that due to simple site exclusion), a tendency for ligands to bind so that they leave vacant lattice sites between nearest neighbors. Some binding systems show a high degree of positive binding cooperativity. For example, the gene 32 protein of bacteriophage T4 binds to single-stranded nucleic acids with a cooperativity ω ranging from the hundreds to the thousands [2]. With such high cooperativity there will be a strong tendency to cover the nucleic acid completely with protein,

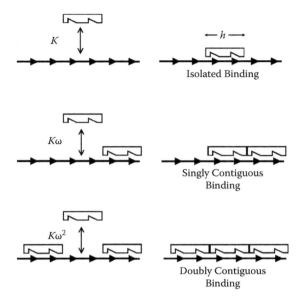

FIGURE 5.2 Parameters for the McGhee-von Hippel model [1]. K is the binding constant for an isolated ligand, ω is the cooperativity factor. Here the site size h is 2.

leaving no open or exposed stretches. This may be advantageous, for example, in protecting exposed single-stranded DNA from attack by nucleases.

It is not possible to have negative values for ω. This is because ω corresponds to a factor in the binding constant, so it must be nonnegative. However, the *free energy change* of the cooperative interaction may be either positive or negative. Values of ω less than 1 correspond to a positive (unfavorable) ΔG, as can be seen by considering the relation between equilibrium constants and free energy changes.

McGhee and von Hippel presented their results for two versions of this model in a form suitable for a Scatchard-type plot. While the Scatchard-type plotting format may not be the most desirable one for a quantitative analysis of the data, it does have the advantage of emphasizing any positive binding cooperativity that is present, and this is certainly an important aspect of binding to linear lattices. For ligands that bind without any cooperative interactions, positive or negative (apart from the effects of site exclusion), McGhee and von Hippel derived the important formula

$$\frac{<v>}{[\text{L}]} = K(1 - h<v>)\left[\frac{1 - h<v>}{1 - (h-1)<v>}\right]^{h-1} \tag{5.7}$$

Scatchard-type plots for different values of h are shown in Figure 5.3A. The binding constant K has been set equal to 1 M^{-1}, as reflected in the intercept on the y-axis. Notice that the intercept on the y-axis will be the intrinsic binding constant K, not K/h as might be expected for the usual Scatchard analysis. Also, the intercept on the x-axis will be at $<v> = 1/h$, corresponding to complete saturation of the binding lattice.

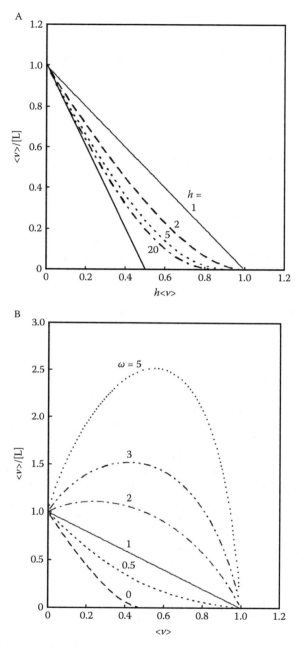

FIGURE 5.3 Calculated Scatchard-type plots (with $<v>/[L]$ as a function of $h<v>$) for the McGhee-von Hippel model of large ligands binding to a linear lattice. A. The noncooperative binding of ligands of different site sizes, with h (in descending order) equal to 1, 2, 5, and 20. The solid line represents an extrapolated tangent to the curve for $h = 20$; its x-axis intercept is at $h < v > = 0.5$. B. The cooperative binding of a ligand with $h = 1$, for $K = 1$ M^{-1}, with ω ranging from 0 to 5.

It is apparent that for $h = 1$ the equation reduces to the simpler original Scatchard form. However, for $h \geq 2$ the second factor in brackets on the right is always less than unity. Furthermore, it is a function of both h and $< v >$. This implies that in the resulting Scatchard-type plot of $< v >/[L]$ versus $< v >$, the curve for the system with "large" ligands will be curved, not a straight line. Also, the curve for the large ligand system will fall below the curve for the $h = 1$ system.

For moderately large ligands the plot becomes very shallow as it approaches the horizontal axis, and for ligands having having $h \geq 5$ it very rapidly becomes quite flat at any appreciable site binding density. The physical reason for the flatness of the curve in this region is the accumulation of numerous small gaps of fewer than h lattice sites between the bound ligands. As a practical matter, this shape of the curve makes it very hard to determine the site size reliably for these larger ligands, because it is very difficult to extrapolate the binding curve to reach the proper intercept. For lattice systems, this is a particularly pertinent reason to bypass the Scatchard plot when one wants a truly quantitative analysis.

The second equation derived by McGhee and von Hippel describes the binding of large, cooperatively interacting ligands. The result is

$$\frac{< v >}{[L]} = K(1 - h < v >)$$

$$\times \left[\frac{(2\omega - 1)(1 - h < v >) + < v > -R}{2(\omega - 1)(1 - h < v >)} \right]^{h-1} \left[\frac{1 - (h+1) < v > +R}{2(1 - h < v >)} \right]^2 \qquad (5.8)$$

with the quantity R given by

$$R = \sqrt{[1 - (h+1) < v >]^2 + 4\omega < v > (1 - h < v >)} \qquad (5.9)$$

(There is a typographic error in the original paper.)

When $\omega = 1$ the isotherm equation reduces (using L'Hôpital's rule) to the previous equation for noncooperative ligands. If the interaction is anticooperative ($\omega \leq 1$), then the theoretical curves for such a system fall below those for the system with $\omega = 1$ and the same value of h. Further, as ω approaches zero, the above relation approaches that for a system with a site size of $(h + 1)$ but with $\omega = 1$. On the other hand, if the interaction is positively cooperative, then the curve will lie above that for the corresponding noninteracting system with the same value for h. The curve will have a characteristic "humped" shape, with the height of the hump and the binding density at which it occurs being a function of the cooperativity. Figure 5.3B shows the effect of small to moderate degrees of cooperativity on the binding curve. The ligand site size has been set equal to 1 here, to focus attention on the binding cooperativity. Notice the pronounced hump in the curves as the cooperativity increases.

For these cooperative-binding models the intercept on the y-axis is still the binding constant K and the intercept on the x-axis is still $1/h$. As a practical matter, most of the easily obtained binding data will lie in the vicinity of the hump, and it is relatively difficult to obtain reliable data at either very low or very high values for $< v >$. This leads to difficulties in determining the intercepts on both the x and y axes. The height of the hump will be pronounced even for moderate levels of cooperativity. Often enough, determining either or both of the intercepts will require a risky extrapolation from the region of the hump to the axis; the extrapolation line may be steep and there may be an appreciable distance separating the region of reliable data and the point of interception.

En route to the isotherm equations, McGhee and von Hippel also derived expressions for various conditional probabilities that might be of interest, including the chance that an unoccupied site would be found next to one end or the other of a ligand, that a free site would be found next to another free site, and so on. These conditional probabilities can be combined to deduce the distribution of clusters of occupied sites, of vacant lattice sites, the average length of a cluster of adjacently bound ligands, etc.

For example, the conditional probability that the left end of one ligand lies to the immediate right of a second ligand (a probability denoted $b_h b_l$ in the notation of McGhee and von Hippel) is

$$b_h b_1 = \frac{1 - (h - 2\omega + 1) <v> - R}{2 <v> (\omega - 1)} \qquad (5.10)$$

for cooperative binding. Using this, the fraction F_c of bound ligand found in clusters of c ligands can be shown to be

$$F_c = c(1 - b_h b_1)^2 (b_h b_1)^{c-1} \qquad (5.11)$$

Also, the average length of a cluster of adjacently bound ligands is

$$\text{Average cluster length} = \frac{1 - (h - 1) <v>}{1 - h <v>} \qquad (5.12)$$

in the case of noncooperative binding, and for cooperative binding it is

$$\text{Average cluster length} = \frac{2 <v> (\omega - 1)}{(h - 1) <v> - 1 + R} \qquad (5.13)$$

These statistical quantities may be experimentally accessible for some systems; for example, electron microscopy of DNA-protein complexes can give estimates of cluster sizes.

Scatchard-type plots for anticooperative systems ($\omega < 1$) and for noncooperative systems in which $h > 1$ are both concave upward, and so with these plots it is difficult to distinguish anticooperativity (deriving, for example, from long-range repulsions between ligands) from simple noncooperativity combined with site exclusion. Also, if there is positive cooperativity but it is weak ($\omega < 5$, say), then the hump in the plot will be small and it will be difficult to distinguish this case from simple noncooperative binding.

An alternative representation of the data is a semilogarithmic plot of the degree of lattice saturation as a function of free ligand concentration. Figure 5.4 shows the

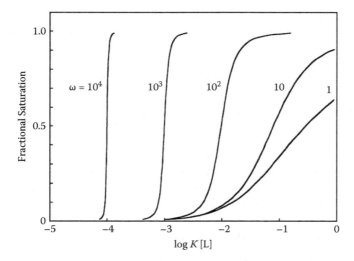

FIGURE 5.4 Semilogarithmic plots of theoretical binding curves (fractional saturation as a function of the product $K[L]$) for the cooperative binding of a ligand of $h = 8$, with ω ranging from 1 to 10^4.

effect of binding cooperativity (ω ranging from 1 to 10^4) on the binding curve for a large ligand. Increasing cooperativity at constant intrinsic affinity (constant K) shifts the midpoint of the curve to lower and lower concentrations of free ligand, while increasing the steepness of the curve. At high cooperativities the binding approaches all-or-none behavior and the lattice becomes saturated with ligand over a very small range of free ligand concentrations. This should be compared to the behavior of the binding curve for $\omega = 1$, in which it is quite difficult to achieve lattice saturation with a large ligand. As discussed earlier, the semilogarithmic format will more reliably show the approach to binding saturation than will the Scatchard plot. Still, even with the semilogarithmic format there remains a difficult problem in nonlinear curve fitting for extraction of model parameters.

McGhee and von Hippel gave two separate formulas, one for the simple noncooperative binding of large ligands ($\omega = 1$), and one for the cooperative (or anticooperative) binding of large ligands ($\omega > 1$ or $0 < \omega < 1$). Unfortunately, if the binding is noncooperative or nearly so, these formulas present certain computational difficulties when fitting data. As ω approaches unity the factor of ($\omega - 1$) in the denominator of the formulas will cause a numerical divergence. While the overall formula does indeed reduce to the proper algebraic form as ω approaches unity, computer programs have difficulty in handling the apparent numerical divergences.

It would be very useful to have a single formula that would apply for all values of the cooperativity parameter, from zero to infinity. Bujalowski et al. [3] have derived just such a formula:

$$\rho = \frac{K[L](1 + \rho + R^*)^2}{4[1 - (1 - \rho - R^*)(2\omega)^{-1}]^{h+1}} \tag{5.14}$$

which they obtained using a different mathematical analysis of the same lattice-binding model. The quantity ρ, the ratio of bound ligands to free lattice residues, is defined by $\rho = <v>/(1 - h <v>)$ The quantities R^* and R are defined by

$$R^* = R / (1 - h < v >) \quad \text{and} \quad R = \sqrt{[1 - (h+1) < v >]^2 + 4\omega < v > (1 - h < v >)} \quad (5.15)$$

The generalized binding equation can be rearranged to the form used by McGhee and von Hippel, and it reduces to the correct form when the cooperativity is set equal to unity (noncooperative binding).

The main advantage of the formulation in terms of ρ is that one equation suffices for all values of ω between zero and infinity. This formulation is a better computational choice when fitting model isotherm equations to experimental data, if the cooperativity ω is one of the parameters being sought.

5.1.3 Extensions of the Model: Oriented Lattices and Ligands

The McGhee-von Hippel model has only three parameters, one of its major virtues. The model, however, does not treat a number of situations that might be encountered with real binding systems. For example, an asymmetric ligand might bind in either of two orientations onto the lattice, with the binding affinity depending on the orientation. Also, the cooperative interaction between two neighboring ligands might depend on their relative orientation. Perhaps a head-to-tail orientation of neighbors would be favored over an arrangement in which two tails point toward one another. This sort of directional binding might arise with biopolymers like DNA, RNA, or proteins, in which the polymer chain has a definite directionality associated with the covalent bonds linking the units.

Another variation on the basic model centers on the number of states available to the lattice sites. The basic McGhee-von Hippel model allows lattice sites to exist in one of two states, unoccupied or occupied, but what if a local conformational change in the lattice were permitted, and due allowance made for changes in cooperativity? One further modification would be to allow for a secondary, piggy-back ligand that modulates the binding of the primary ligand. The theory to describe these and other variations on the basic model has thus continued to develop [3–10]. Figure 5.5 presents a selection of these models.

The statistical weights assigned to the three-lattice site states are u, a, and b. Table 5.1 (adapted from Bujalowski et al. [3]) shows the formulas to be used in computing these statistical weights and that of the equilibrium constant K_0. The table also shows the formula for the intrinsic binding constant K, as determined by extrapolation in a Scatchard-type plot to zero-site binding density. Although three states for a lattice site are used, up to *four* different cooperativity parameters are now needed.

To deduce expressions for the binding equation for the general three-state model, Bujalowski et al. approached the problem using the sequence-generating function of Lifson [11]. The basics of this powerful and general tool are given in the Appendix for the interested reader. A key step here is that the partition function Z_N for

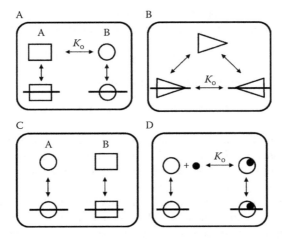

FIGURE 5.5 Selected schemes for various three-state lattice-binding models. Different binding affinities and site sizes are assumed for each binding mode illustrated. See Table 5.1 for model parameters. A. Ligand switches with equilibrium constant K_0 between two conformations with different affinities and cooperativities. B. Ligand binds in two different orientations, with different affinities and cooperativities, on an oriented lattice. C. Two different ligand species compete for binding to the lattice. D. "Piggyback" binding of a secondary ligand (symbol •) to the primary ligand. With permission of John Wiley & Sons, Inc. Adapted from *Biopolymers* Vol. 28, W. Bujalowski, T.M. Lohman, and C.F. Anderson, "On the cooperative binding of large ligands to a one-dimensional homogeneous lattice: The generalized three-state model," pages 1637–1643. Copyright 1989 John Wiley & Sons.

TABLE 5.1
Parameters for Selected Three-State Lattice Models

	Statistical Weights			"Conformational" Equilibrium Constant	"Intrinsic" Binding Constant
Model	u	a	b	K_0	K_i
A	1	$K_A[L_A]$	$K_B[L_B]$	$[L_B]/[L_A]$	$(K_A + K_B K_0)/(1 + K_0)$
B	1	$K_A[L_A]$	$K_B[L_B]$	K_B/K_A	$K_A(1 + K_0)$
C	1	$K_A[L_A]$	$K_B[L_B]$	0	K_A or K_B
D	1	$K_A[L_A]$	$K_B[L_B]$	$[L_B]/([L_A][R])$	$(K_A + K_B K_0[R])/(1 + K_0[R])^\dagger$

Adapted from Bujalowski et al. [3].

Note: K_x denotes the binding constant for species x, in the limit of isolated lattice binding of species x; K_i is the constant determined by extrapolation to the limit of zero site binding density, for the full system.

†R denotes piggy-back ligand species; equation holds only for $[R] \gg [L]$.

a lattice of N sites can be expressed simply as $Z_N = x^N{}_1$, where x_1 is the largest root of the model's secular equation, a complicated polynomial expression:

$$x^{h+m+1} - x^{h+m} - \omega_2 b x^{h+1} - \omega_1 a x^{m+1} + (\omega_2 - 1)b x^h$$

$$+ (\omega_1 - 1)a x^m + (\omega_1 \omega_2 - \omega_3 \omega_4)abx$$

$$+ (\omega_1 + \omega_2 - \omega_3 - \omega_4 - \omega_1 \omega_2 + \omega_3 \omega_4)ab = 0$$

$$(5.16)$$

Here x is a dummy variable whose numerical value is determined by satisfying the above equation. Also, h is the site size for binding to lattice sites in one state, and m is the site size for binding to lattice sites in a second, different state (m is used here to indicate a site size, in which Bujalowski et al. [3] have used k; this should avoid confusion of a site size parameter with a binding constant). Formulas for the statistical weights a and b are given in Table 5.1 for various three-state models.

The next step toward finding an expression for the binding equation was to use the partial derivative of Z_N to express the site binding density as a function of free ligand activity:

$$<v> = \frac{1}{N}\left(\frac{\partial \ln Z_N}{\partial \ln a_L}\right) = \left(\frac{\partial \ln x}{\partial \ln a_L}\right)_{x=x_1}$$

$$(5.17)$$

In general, it is not possible to solve the secular equation analytically for its roots (only if $[h + m + 1]$ is 5 or less is this algebraically possible), but it can be solved numerically. By picking values for the free ligand concentration, the site sizes, and the cooperativities, the secular equation can then be solved numerically for a particular value of its largest root x_1. By repeating the process for a slightly different free ligand concentration, it is then possible to form a numerical derivative of x_1 with respect to free ligand concentration, and hence calculate the site binding density v for that particular free ligand concentration. With many repetitions of the overall procedure one can (eventually) generate the binding curve expected for those particular parameter values. To compare the curve with experimental data, one should systematically vary the parameters to achieve the best statistical fit. But with eight parameters to vary, the search through parameter space for the "best" fit can be quite time consuming and there is no guarantee that the search will converge on a single set of parameter values.

Of course, by setting various of these cooperativities equal to one another or equal to unity, it is possible to obtain simpler (though less comprehensive) models. For example, by setting all four cooperativities equal to a single common value, and by setting the statistical weight b equal to zero, one can obtain a secular equation for the original McGhee-von Hippel model:

$$x^{h+1} - x^h - a\omega x + (\omega - 1)a = 0$$

$$(5.18)$$

where $a = K[L]$. Bujalowski et al. show how this simplified secular equation leads to the Scatchard plot formulation given by McGhee and von Hippel.

5.1.4 SITE-EXCLUSION IN DNA-POLYAMINE BINDING

The interactions of polyamines and oligopeptides with DNA can be viewed as prototypes of the more complicated protein-nucleic acid interactions involved in gene expression, DNA replication, recombination, and repair. The binding mechanisms for polyamines and oligopeptides are usually simpler and easier to analyze, while still giving considerable insight into the more complicated systems operating in vivo.

As model ligands for the effects of ionic charge on binding, Braunlin et al. [12] studied the binding of the polyamines spermine (with a valence of +4), spermidine (+3), and putrescine (+2) to double-stranded DNA. Using equilibrium dialysis with ^{14}C-labeled polyamines, they obtained isotherms that they analyzed using the McGhee-von Hippel model for noncooperative binding. They also carried out competition binding experiments, competing both Mg^{2+} and putrescine against spermine and spermidine, and analyzing these isotherms with an extended version of the basic McGhee-von Hippel equations that allowed for competitive binding of more than one ligand species [5]. In all cases the binding was sensitive to salt concentration. Plots of log K_{obs} versus log [Na^+] were linear, and extrapolated intercepts at 1 M NaCl were numerically quite small, as would be expected for a binding interaction driven primarily by electrostatic attraction. Nonlinear least-squares fitting using the model equations gave values for the ligand site sizes that were in reasonable accord with the relative molecular sizes of these polyamines. The values for the site size h did vary with salt concentration; e.g., for spermine (ionic charge of +4) h ranged from 3.6 to 5.4, for spermidine (+3 charge) h ranged from 2.1 to 3.7, and for putrescine (+2 charge) h ranged from 1.7 to 2.5.

Employing equilibrium dialysis, Latt and Sober [13] studied oligolysine binding to two different synthetic double-stranded RNAs, poly(rI) • poly(rC) and poly(rA) • poly(rU). Their ligands were a series of oligolysines that carried a fluorescent dinitrophenyl group as a label at one end. The positive charges on the ε-amino groups of the side chains of the lysine residues were expected to provide substantial attraction of the oligopeptide for the negatively charged phosphate backbone of the nucleic acids. Binding saturation for either polynucleotide species was not reached; in particular, the oligolysines formed a precipitate with poly(rA) • poly(rU) well before reaching binding saturation.

The binding data were analyzed using a model in which an oligopeptide of n residues was assumed to bind to n residues on the polynucleotide in such a way as to occlude $a \geq n$ lattice sites. These groups, each of a lattice sites, were assumed to be distributed over a polynucleotide of N total sites, without any overlapping of sites. Each group of a adjacent lattice sites was treated as an independent entity, and the groups and vacant sites were distributed according to binomial statistics. (This work was, of course, done years before the development of the McGhee-von Hippel model, and the procedure unfortunately miscounts the number of lattice sites available for binding.) With K the intrinsic binding constant, and γ a constant that is determined by the value of a, Latt and Sober proposed an approximate binding equation for this model:

$$Y = \frac{aK[L]}{1+(a+\gamma)K[L]} \tag{5.19}$$

This can be recast for plotting data in either the Scatchard-type format or the double-reciprocal format to obtain the binding constant, the number of occluded sites per ligand, and the parameter γ. However, Latt and Sober reported only values for the binding constant and not for the other two parameters.

McGhee and von Hippel [1] reanalyzed the data using their own model for site exclusion, without positive cooperativity. Scatchard-type plots of the data showed the concave-up curvature characteristic of simple site exclusion; there were no humps in the plots, such as would appear if there were positive cooperativity present. (However, a minor amount of positive cooperativity might well have gone undetected; a small degree of positive cooperativity can be masked by the negative effects of site exclusion.) Isotherms for oligolysines with five, six, and seven residues were clearly distinguishable from one another.

The experimental data do not approach the region of binding saturation, and (as expected from the form of the theoretical isotherms) it is difficult to determine the x-axis intercept by inspection of the Scatchard-type plot. McGhee and von Hippel noted that to achieve 95% saturation with one of these systems, one would have to use free ligand concentrations in excess of 0.5 M, an experimental impossibility with these ligands due to issues of solubility.

The data were fit numerically with the noncooperative-binding version of the McGhee-von Hippel model, varying the site size and the binding constant to achieve the best fit. Curve fitting gave estimates for the binding constant K that were about 25% larger than those estimated by Latt and Sober. Interestingly, the site sizes found by curve fitting were consistently larger (by 1.2, on the average) than the number of amino acid residues in the oligopeptides. If one ε-amino group neutralized the charge on one phosphate group, then the site size might be expected to numerically equal the number of amino acid side chains present in the oligopeptide. The apparent "excess length" in the site size might be due to the bulky dinitrophenyl group at the end of the oligopeptide, or it might be caused by coulombic repulsion between the positively charged ligands. Regardless of this discrepancy, however, both the binding constant and site size values are physically reasonable, and the theoretical model provides a good fit to the data.

5.1.5 SITE-EXCLUSION AND POSITIVE COOPERATIVITY IN DNA-PROTEIN BINDING

The binding simulations in Section 5.1.2 showed that the negative effects of site exclusion can be offset by positive cooperative interactions between neighboring ligands on the lattice. This applies even to very large ligands such as entire proteins, where there may be only a modest intrinsic binding affinity, perhaps on the order of $K = 10^3$ M^{-1}. A positive cooperative interaction between bound proteins with a free energy change of a few kilocalories per mole will produce a system that saturates the lattice quite readily. This magnitude of free energy change corresponds to ω on the order of 100, and it could be readily generated by just one or two weak attractive interactions between the adjacent ligands. The combination of a large site size, a weak intrinsic affinity, and a moderate to high degree of binding cooperativity is in fact seen with several proteins that bind to DNA or RNA, where for one reason or another (perhaps for protection against nuclease attack, or for compaction for storage) it is important to cover the nucleic acid completely with a protein coat.

An example is the gene 5 protein from the filamentous bacteriophages f1, fd, and M13. This protein binds preferentially to single-stranded DNA over double-stranded DNA. The binding is highly cooperative, which is important for the role that the protein plays in the replication of the virus. When the protein has accumulated to a certain critical level in an infected cell, it triggers a change from synthesis of the replicative form of the virus (double-stranded circles) to single-stranded DNA synthesis, doing this by covering the newly synthesized single-stranded DNA with contiguous copies of the protein.

Alma et al. [14] used binding-induced quenching of the protein's intrinsic fluorescence to follow the binding of gene 5 protein to various single-stranded polynucleotides. Panel A of Figure 5.6 is a Scatchard-type plot of the binding to poly(dA). The high degree of binding cooperativity is evident in the pronounced hump in this plot. The plot also illustrates some of the difficulty one would have in deducing a site size or an intrinsic binding constant from such a plot, since the binding curve's intercepts on the x and y axes are far from being precisely determined. Panel B of Figure 5.6 compares binding to poly(dA) across several different salt concentrations, and to M13 DNA. The data have been fit with the McGhee-von Hippel model for cooperative binding of large ligands. According to the model, the reciprocal of the free protein concentration at the midpoint of the titration will equal $K\omega$. The plot in panel B shows that it is relatively easy to extract this value. However, detailed curve fitting must be used to resolve K from ω, and to extract the site size h. The apparent site size for gene 5 protein is about four nucleotides, and the cooperativity ω ranges from 50 to 140 for binding to poly(dA). Similar values are found for this protein binding to single-stranded M13 DNA. The intrinsic site binding affinity K is relatively modest, on the order of 10^3 to 10^4 M^{-1}. Salt has a strong effect on the intrinsic binding, but rather little effect on the cooperativity. This indicates that the cooperativity likely does not involve electrostatic interactions, though electrostatics seem to be important for the initial binding of the protein to the nucleic acid lattice.

These parameter values are consistent with a model of viral replication in which the gene 5 protein displaces the SSB protein of *E. coli*. The SSB protein is also a single-stranded DNA binding protein that has a greater affinity (larger K) than the gene 5 protein, but also a much larger site size (approximately 35 nucleotides). By virtue of the high binding cooperativity of the gene 5 protein, the average nucleotide affinity of the gene 5 protein over a run of 35 nucleotides, where eight to nine monomers are bound contiguously, is much greater than that for a single SSB protein over the same run of nucleotides. This is the reason that the gene 5 protein is able to displace the SSB protein from the single-stranded viral DNA and so initiate replication of the single-stranded form necessary for packaging of the virus. The combination of weak affinity and high cooperativity may be a general mechanism for the replacement of one DNA binding protein by another.

5.1.6 SITE-EXCLUSION AND LATTICE CONFORMATIONAL CHANGE

Certain polyaromatic dyes and drugs bind to DNA via *intercalation* of the ligand between the stacked base pairs of the helix of the nucleic acid (see Figure 1.11). Lerman [15] first proposed this mode of binding to describe the interaction of DNA

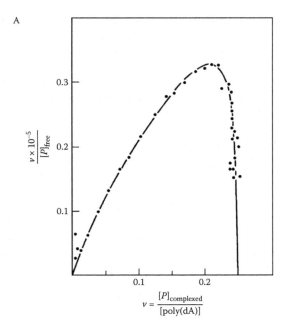

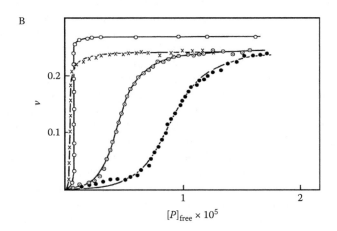

FIGURE 5.6 Cooperative binding of the gene 5 protein of bacteriophage M13, to single-stranded poly(dA). A. Scatchard-type plot; $[Na^+] = 0.2$ M. The solid line is a statistical best fit using the full McGhee-von Hippel model of cooperative binding, yielding the values $h = 4.1$, $\omega = 180$, and $K\,\omega = 2.0 \times 10^5$ M^{-1}. Note the use of the symbol v for the site binding density. B. Site binding density as a function of free protein concentration $[P]_{free}$, for various salt concentrations. (•) Binding to poly (dA), $[Na^+] = 0.22$ M; (◉) binding to poly (dA), $[Na^+] = 0.185$ M; (×) binding to poly (dA), $[Na^+] = 0.11$ M; (○) binding to M13 DNA, $[Na^+] = 0.20$ M. From *Journal of Molecular Biology* Vol. 163, N.C. Alma, et al., "Fluorescence studies of the complex formation between the gene 5 protein of bacteriophage M13 and polynucleotides," pages 47–62. Copyright 1983, with permission from Elsevier.

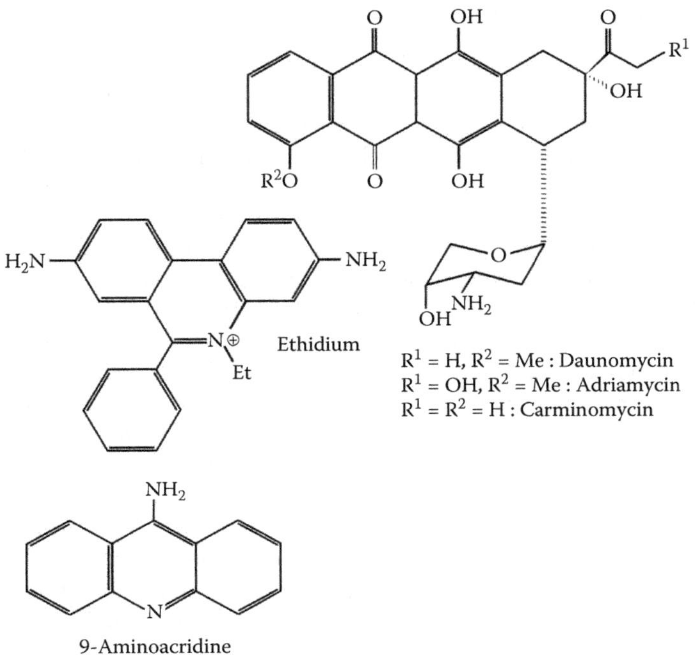

FIGURE 5.7 Ethidium and other common intercalative agents.

with acridine derivatives, but it has since become apparent that many other planar aromatic dye and drug molecules bind to nucleic acids this way as well. (Figure 5.7 shows some typical intercalators.) Intercalation leaves intact the hydrogen bonding between bases in double-stranded DNA, but requires untwisting of the double helix to permit formation of a suitable space between the paired bases for insertion of the drug molecule. As a result, across the intercalation site the neighboring base pairs are about twice as far apart as normal. This extends the length of the DNA molecule. The DNA backbone is distorted, mainly through alterations in the puckering of the ribose moiety. The local distortion of the DNA helix tends to block any further intercalation at the two base pairs immediately adjacent. In effect, ligand occupation at one site will exclude neighboring base pairs to either side from accepting another intercalator ligand.

The theoretical effect exerted by neighbor exclusion on the binding curve was described early on by Donald Crothers [16] and later elaborated by Bresloff and Crothers [17,18]. It can, of course, be treated crudely by the basic McGhee-von Hippel model, but this would ignore any differences in binding cooperativity as a function of lattice conformation (e.g., unwinding of the DNA by the intercalating agent). The general three-state model can treat the combined effects of changes in the lattice conformation and cooperativity and site exclusion [3,4], and with the power of today's desktop computers this would now be the preferred choice for use in curve fitting. However, much of the data in the literature has been analyzed using simpler models. An example is that of ethidium bromide binding to DNA.

5.1.6.1 Ethidium/DNA Interactions

Ethidium is a cationic dye that is widely used in studies of nucleic acids. Probably its most common laboratory use is as a fluorescent stain to detect DNA in gel electrophoresis experiments. In common with other intercalating compounds, ethidium has a wide range of biochemical effects, including inhibition of various DNA-dependent enzymes such as RNA polymerase, and induction of frameshift mutagenesis.

Both kinetic and equilibrium measurements indicate that ethidium binds to double-stranded DNA in two modes. The first and stronger mode is by intercalation; the second and weaker mode is nonintercalative, probably with the negatively charged DNA backbone acting as a template for aggregation of the cationic dye [19]. Intercalation is definitely the more interesting of the two modes of binding.

Ethidium has a weak preference for pyrimidine-(3′-5′)-purine sequences [18,20]. The binding constants vary with the salt concentration, with $\partial \ln K / \partial \ln [Na^+]$ ranging from -0.8 to -1.1, depending on the DNA species. Striking differences in the entropy change ΔS are seen with binding to homopolymer duplexes, compared to the alternating duplex copolymers, and yet again with a change in the type of backbone sugar. This may be connected to differences in the relative hydration of the various polynucleotides [20]. Differential scanning calorimetry of the dissociation of ethidium from the two types of AT polymers shows that differences between the alternating copolymers and homopolymers disappear at 92°C. Thermal agitation presumably "levels" the hydration of the two types of polymer. Compared to behavior at room temperature, the binding at high temperature is more exothermic ($\Delta H°$ is more negative), and the entropy change is very much more negative, rather closer to what one might expect from a pure bimolecular association [20].

An interesting question in the binding of ethidium to DNA is whether the binding shows any positive cooperativity. Early studies reported anticooperative binding isotherms that were consistent with the neighbor exclusion model [17]. The crystal structure of the drug intercalated into an oligonucleotide provided a structural rationalization of binding site exclusion for a site size of 2 [21]. However, there were little or no data at very low site binding densities, that is, below $< v > = 0.05$, and it is in this region of $< v >$ values that (weak to moderate) positive cooperativity might appear.

In studying ethidium binding to various synthetic polynucleotides, Bresloff and Crothers (see Figures 5.8 and 5.9) found some evidence for positive cooperativity [18]. They analyzed part of their data using a version of the neighbor exclusion model equivalent to that of McGhee and von Hippel for cooperative binding of large ligands. The binding to poly d(AT) • poly d(AT), for example, could be fit with a site size of 2 (consistent with the basic neighbor exclusion model of intercalation) and a cooperativity of 2.3 (only weak positive cooperativity).

Binding to poly d(IC) • poly d(IC) could not be fit by any simple model of site exclusion with cooperativity, and here they used instead a more general model involving cooperative binding to two different conformations of the nucleic acid lattice. This model employs nine different parameters or statistical weights and is mathematically equivalent to one version of the three-state model as given by Bujalowski et al. [3]. Binding to each of the two nucleic acid conformations is characterized by a different

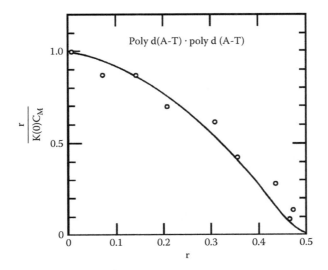

FIGURE 5.8 Ethidium binding to poly d(A-T) • poly d(A-T). Solid curve represents a statistical fit yielding a site size of 2, with $K = 2.9 \times 10^4$ M^{-1}, $\omega = 2.3$. Note the use of the symbol r for the site binding density, and the symbol K(0) for the binding constant of an isolated ligand. From *Biochemistry* Vol. 20, J.L. Bresloff and D.M. Crothers, "Equilibrium studies of ethidium-polynucleotide interactions," pages 3547–3553, with permission. Copyright 1981 American Chemical Society.

binding constant, a different site size, and a different cooperativity parameter. The conformational switching of the nucleic acid is governed by two parameters. Using the notation of Bresloff and Crothers, the first is an equilibrium constant per base pair s for converting a base pair from Form I to Form II, at the junction of a region of Form I and Form II. The second conformational parameter is a nucleation parameter σ representing the difficulty in initiating such a conversion in the middle of a region that is purely Form I. The equilibrium constant for converting a single base pair from Form I to Form II, in the middle of a region that is purely Form I, is $\sigma^2 s$.

For Form I, K_1 was found to be 2.6×10^4 M^{-1}, while K_2 was 1.4×10^4 M^{-1}. The site size was between 2 and 3 for Form I, and was set at 2 for Form II. The binding cooperativity was negligible for Form I ($\omega = 1$), and slightly positive for Form II ($\omega = 3$). Qualitatively, Bresloff and Crothers found that intrinsic binding was weaker to Form II than to Form I, but that the slight positive cooperativity in binding to Form II would compensate for the lesser degree of intrinsic affinity for Form II at higher degrees of binding density.

Winkle et al. [22] also investigated this region of the binding isotherm, finding telltale humps in Scatchard-type plots of their data when they used DNA from *E. coli* (see Figure 5.10). However, they found no evidence of positive cooperativity for ethidium binding to other natural DNAs. The *E. coli* data could be fit using the lattice conformational-change model (in this particular application, ligand-ligand cooperativity was neglected). Unfortunately, the data could also be regarded as consistent

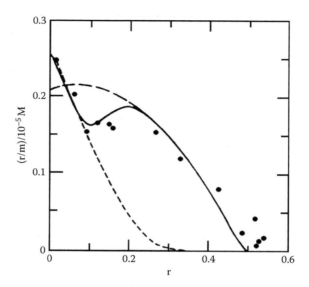

FIGURE 5.9 Fitting a two-state lattice model for ethidium binding to double-helical poly d(I-C) • poly d(I-C). Note the use of r for the site binding density; m represents the molarity of free ligand. The solid line is the isotherm calculated using the following parameters: a site size of 2–3, $K = 2.6 \times 10^4$ M^{-1}, $\omega = 1$ for Form I; a site size of 2, $K = 1.5 \times 10^4$ M^{-1}, $\omega = 3$ for Form II; and conformational switching parameters $s = 0.98$ and $\sigma = 0.01$. From *Biochemistry* Vol. 20, J.L. Bresloff and D.M. Crothers, "Equilibrium studies of ethidium-polynucleotide interactions," pages 3547–3553, with permission. Copyright 1981 American Chemical Society.

with a model in which ethidium binds very cooperatively to a limited region of the *E. coli* DNA, and noncooperatively to the majority of the remaining DNA sites.

The study by Winkle et al. illustrates a difficulty in working with natural DNA preparations. There will be, of course, substantial heterogeneity in the sequence of such DNAs, and so there is always the possibility of heterogeneity in binding affinity as well. A separate complication is that the binding curves for slightly different versions of the general McGhee-von Hippel model tend to lie rather close to one another. Even with synthetic polynucleotides (where presumably the sites are homogeneous in their binding affinity) the question of which of several models best fits the data may not be easily resolved.

5.1.7 PIGGY-BACK BINDING AND DNA-PROTEIN INTERACTIONS

The binding activity of a number of DNA-binding proteins can be modulated by small organic molecules that are bound by the protein but not by the nucleic acid. As mentioned earlier, this kind of piggy-back binding is a common theme in gene regulatory mechanisms. One such regulatory protein with this piggy-back mechanism is the cyclic AMP receptor protein (CRP) of *E. coli*.

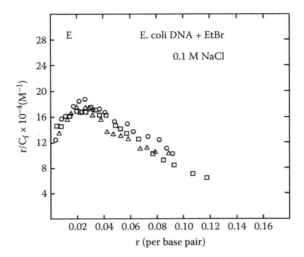

FIGURE 5.10 Cooperativity in ethidium binding to *E. coli* DNA. Note the use of r for the site binding density and C_f for the free ligand concentration. Data shown as (○) and (□) are duplicate experiments; data shown as (△) were obtained with *E. coli* DNA treated with S1 nuclease to remove single-stranded material. From *Nucleic Acids Research* Vol. 10, S.A. Winkle, L.S. Rosenberg, and T.R. Krugh, "On the cooperative and noncooperative binding of ethidium to DNA," pages 8211–8223, by permission of Oxford University Press. Copyright 1982 Oxford University Press.

The CRP-cyclic AMP system has served for years as a model of transcriptional gene regulation [23]. CRP frequently acts as a positive regulator of transcription, and its activity is affected by its binding of cyclic AMP (cAMP). The lactose operon of *E. coli* is one of several operons where piggy-back binding of cAMP to CRP is important. CRP binds to DNA in the region near the *lac* promoter in a ternary complex with cAMP, and it helps to stimulate RNA polymerase action at the *lac* promoter. Since cAMP has negligible affinity for DNA itself, this is an example of piggy-back binding, the cAMP riding on CRP as CRP binds to DNA. The protein can also bind cooperatively to single-stranded or double-stranded DNA nonspecifically, though with a preference for double-stranded DNA [24,25]. cAMP increases the overall affinity for DNA but reduces the cooperativity. As noted above, CRP acts as a positive regulator of transcription in the *lac* system [26] and in many others [23]. However, it can also act as a negative regulator as well; for example, it is a repressor of its own expression [23]. Overall, CRP shows an interesting variety of binding phenomena.

The protein is relatively small and simple, a dimer of identical subunits with a total molecular weight of 45,000. The crystal structure of the protein shows two distinct domains in each of the subunits; one of these contains the binding site for cAMP, the other contains three α-helices that bind to DNA [26,27]. The crystal structure of CRP in complex with DNA shows that CRP induces a sharp 90° bend in DNA when bound to its specific recognition sequence, with implications for its interaction with RNA polymerase [28].

Binding of cAMP by CRP in the *absence* of DNA can be described by the binding polynomial for a cooperative, two-site system:

$$\mathbf{P} = 1 + 2K[L] + \alpha K^2[L]^2, \quad <r> = \frac{2K[L] + 2\alpha K[L]^2}{1 + 2K[L] + \alpha K^2[L]^2} \tag{5.20}$$

where α is a cooperativity parameter for binding of the second ligand. This binding equation was modified slightly by Takahashi et al. [29] in their report, as

$$<r> = \frac{nK[L](1 + 2\alpha K[L]^2)}{1 + 2K[L] + \alpha K^2[L]^2} \tag{5.21}$$

where the parameter n, the number of binding sites, was left as a quantity to be determined by curve fitting. This is unfortunately inconsistent with the presence of the term $\alpha K^2[L]^2$ in the equation, whose presence already implies two identical binding sites on the protein.

In their study, Takahashi et al. used the method of equilibrium dialysis, following the uptake of radioactively labeled cAMP. Isotherms for cAMP-CRP binding showed curvature that depended on the salt concentration used. At low salt the isotherms were concave-up, consistent with negative cooperativity in the uptake of cAMP ($\alpha < 1$), while at high salt the isotherms curved in the opposite sense, consistent with positive binding cooperativity ($\alpha > 1$). The x-axis intercepts were consistent with the presence of two sites per dimer, and the y-axis intercepts indicated an intrinsic binding constant of about $4 - 5 \times 10^4$ M^{-1}, which appeared relatively insensitive to salt concentration. The cooperativity parameter, however, was salt sensitive, varying from 0.6 (at 0.05 M salt) to 2.3 (at 0.60 M salt). Binding of cAMP was nearly noncooperative ($\alpha = 1.1$) at 0.40 M salt.

Using a centrifugation method and the McGhee-von Hippel model of site exclusion for large ligands on a linear lattice, Saxe and Revzin [24] found that the nonspecific binding constant for CRP to DNA was about 10^4 M^{-1} at $0.06 - 0.07$ M salt and 22°C, and the cooperativity (ω) in DNA binding was about 100. The binding constant varied with salt concentration in a way that indicated release of about six monocations as the protein bound to DNA. The addition of cAMP to the system slightly increased the binding to DNA, with only a minor effect on the DNA binding cooperativity. Blazy et al. [30] found similar results, using CD spectroscopy to monitor the binding.

In a later study, Takahashi and colleagues studied binding of cAMP to preformed complexes of CRP with both specific and nonspecific sites on DNA [31]. As experimental methods they used both fluorescence measurements and equilibrium dialysis of labeled cAMP. They used very low DNA binding densities (more than 100-fold molar excess of DNA over protein) to reduce the effects of site exclusion and so to approach the situation of studying isolated DNA-protein complexes (a mean protein cluster size of no more than 1.5). By combining these new results with those from previous work on simple CRP-DNA binding and cAMP-CRP binding, they were able to dissect out the binding constants for CRP to a specific DNA site, for CRP

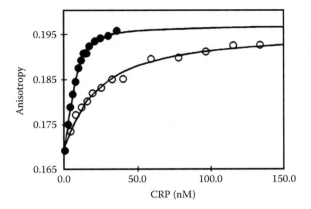

FIGURE 5.11 Binding of CRP to 32-base pair fragment of the *lac* promoter determined by fluorescence anisotropy spectroscopy. (o), 0.5 µM cAMP; (●), 500 µM cAMP. Solid lines represent a best fit for a single-site isotherm. From *Proceedings of the National Academy of Sciences USA* Vol. 87, T. Heyduk and J.C. Lee, "Application of fluorescence energy transfer and polarization to monitor *Escherichia coli* cAMP receptor proteins and lac promoter interaction," pages 1744–1748. Copyright 1990 National Academy of Sciences U.S.A.

molecules carrying either one or two molecules of cAMP. This analysis specifically allowed for the piggy-back binding. They found that binding of a single molecule of cAMP increases CRP site-specific binding to DNA; binding of a second molecule of cAMP enhances the site-specific binding only slightly, compared to the effect of the first cAMP molecule bound. Similar results were obtained by Heyduk and Lee (see Figure 5.11) [32]. These workers used fluorescence energy transfer and fluorescence polarization methods to study the binding of CRP to a 32-base pair fragment of the *lac* promoter, and again the piggy-back model was used to analyze the results. The figure shows the strong influence of cAMP concentration on the CRP-DNA binding.

The quantitation of these binding affinities finds an important application in modeling overall control of the *lac* operon, in particular the influence of nonspecific DNA binding and the overall sensitivity to levels of cAMP [31]. Furthermore, the cooperativity of CRP binding to DNA may be important in other operons (e.g., that for galactose) where there are multiple binding sites for CRP [23].

5.2 BINDING TO TWO-DIMENSIONAL LATTICES

Just as one-dimensional lattice models are useful in interpreting results for binding to linear polymers, two-dimensional lattice models can be useful in modeling ligand binding to the flat surfaces of membranes, crystals, catalysts, and other patterned planar systems. As before, the simplest case to treat is that of independent binding of the ligands, in which there is no interaction of a bound ligand with its neighbors. And once again, this leads to the Langmuir binding isotherm equation. But two-dimensional lattices offer considerably greater complexity than do the one-dimensional kind. With two-dimensional arrays, there might be interesting effects upon changing the geometrical type of the array, e.g., rectangular versus hexagonal arrays, especially when considering cooperative interactions and site-exclusion effects by large ligands.

Unfortunately, the theory for binding to two-dimensional lattices quickly runs into great mathematical difficulties that one does not encounter with one-dimensional lattices. Physicists have long been interested in the corresponding two-dimensional spin lattice and lattice gas problems as models for the behavior of solid matter, and many different approximation schemes have been employed to calculate the partition function for such systems. Exact results with closed-form expressions for such systems are few and far between. There is a useful approximate scheme, however, that has given qualitative insight into the behavior of binding to two-dimensional lattices.

5.2.1 HEURISTIC TREATMENT OF SITE EXCLUSION IN TWO DIMENSIONS

5.2.1.1 The Stankowski Model

Using an approach involving conditional probabilities, rather like that of McGhee and von Hippel [1], Stankowski developed an approximate theory for the binding of large ligands to a two-dimensional lattice [33–35]. He was able to treat the effects of ligand size and shape, and of cooperative interactions among ligands as well. The formulas he derived are relatively simple and they bear a strong resemblance to those derived by McGhee and von Hippel for binding to a one-dimensional lattice. The theory can be applied directly to real systems to gain qualitative insights into the binding.

Stankowski started with a regular array of identical lattice sites on a surface large enough so that edge effects could be neglected. To allow for different lattice geometries giving different numbers of nearest neighbors for the lattices sites, Stankowski used a coordination number z, the number of nearest neighbors surrounding each site. An incoming ligand would cover h sites in a defined spatial arrangement (Stankowski's notation here and below has been changed to conform with conventions used in this book); the main requirement is that the covered sites are contiguous or connected. The arrangement of the h sites might be that of a long straight chain, a closed polygon, or some other (possibly irregular) shape. See Figure 5.12.

Stankowski denoted the fraction of free lattice sites as X_1, and the fraction of bound sites as X_2 (X_2 is also the degree of saturation of the lattice). The site binding density was defined as $<v> = X_2/h$, as in the McGhee-von Hippel theory. To describe the dependence of binding on the shape of the ligand a new parameter, β, was introduced. This parameter is the ratio of "interior" to "exterior" nearest-neighbor lattice site contacts, for an isolated ligand. For large ligands with compact ligand shapes β will be greater than 1, but for elongated shapes β will be less than 1. A related parameter, λ, was defined by $\lambda = 1 / (\beta + 1)$.

Next, Stankowski defined $X_{(h)}$ as the mole fraction of lattice sites that were available for binding an h-mer ligand. This could also be interpreted as the probability of finding h free sites in an arrangement to fit the shape of the ligand. The law of mass action for binding an h-mer ligand could then be written as

$$k[\text{L}] = \frac{<v>}{X_h} \tag{5.22}$$

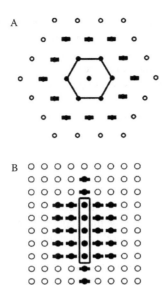

FIGURE 5.12 Ligand shape and lattice geometry become important in binding to surfaces. Excluded region about the ligand is indicated by the filled lattice points. A. Site exclusion by a hexagon ($h = 7$) on a hexagonal lattice, resulting in an excluded area of 12 lattice points. B. Site exclusion by a linear rod ($h = 5$) on a square lattice, resulting in an excluded area of 24 lattice points.

Using conditional probability arguments, Stankowski first derived a formula for $X_{(h)}$, for the binding of a large *linear* ligand:

$$X_{(h)} = \rho(1 - X_2)\left[\frac{1 - X_2}{1 - (1 - \lambda)X_2}\right]^{h-1} \tag{5.23}$$

The quantity ρ is the coordination number divided by the symmetry number of the ligand; this parameter is the number of distinct ways a ligand can be arranged on the lattice if one of its points of contact to the lattice is held fixed.

Defining $k_{eff} = k\rho$ and applying the formula for $X_{(h)}$ to the law of mass action, Stankowski obtained

$$k_{eff}[\text{L}] = \frac{X_2}{h(1 - X_2)}\left(\frac{1 - (1 - \lambda)X_2}{1 - X_2}\right)^{h-1} \tag{5.24}$$

which could be converted to a form suitable for a Scatchard-type plot:

$$\frac{<v>}{[\text{L}]} = k_{eff}(1 - h<v>)\left(\frac{1 - h<v>}{1 - (1 - \lambda)h<v>}\right)^{h-1} \tag{5.25}$$

This was the principal result for noncooperative binding of large *linear* ligands to a two-dimensional lattice. Notice that binding of an h-mer ligand to a linear lattice is a special case of this last relation. For a linear lattice, the coordination number z of a site must equal 2, so that the parameter λ must be equal to $1/h$. With this substitution one recovers the McGhee-von Hippel result for noncooperative binding of an h-mer to a long linear lattice.

For more general ligand shapes, Stankowski first defined an excluded-area parameter α as

$$\alpha = \frac{\text{excluded area}}{h} \tag{5.26}$$

(This should not be confused with the earlier use of the symbol α as a cooperativity factor.) The excluded area would depend on both the type of lattice and the shape of the ligand; various possibilities for ligands with high symmetry are given in Table 5.2.

Then, using some ideas from Andrews' treatment of two-dimensional fluids [36,37], Stankowski found a model isotherm expression:

$$k[\text{L}] = \left(\frac{<v>}{1 - h<v>} \right) e^{-\frac{\alpha h<v>}{1 - <a><v>}} \tag{5.27}$$

The parameter $<a>$ was the average area occupied by a ligand when the ligands are packed close enough to avoid molecular-size vacancies on the lattice. To a good approximation this could be taken as the average area occupied by a ligand when the lattice is saturated. For regular polygons capable of completely covering the surface (e.g., squares, hexagons, etc.) $<a>$ could be simply set equal to h, the number of sites occupied by the ligand. With appropriate values of the excluded-area parameter

---

TABLE 5.2
Stankowski Model for Large Ligands on Two-Dimensional Lattices

Lattice Type	Coordination Number	Ligand Shape	Excluded Area Parameter	Remarks
Square	4	Linear rod length h	$(h^2 - 1)/2h$	
		Square side length a	$(3a - 1)(a - 1)/h$	$h = a^2$
		Rectangle side lengths a and b	$[(a + b)(a + b - 4) + 2(h + 1)]/2h$	$h = ab$
Hexagonal	6	Linear rod length h	$(2h^2 - h - 1)/3h$	
		Hexagon k concentric shells around center	$3k(3k +1)/h$	$h = 1 + 3k(k + 1)$ $k = [- 1 + (1 + 4(h - 1)/3]^{1/2}/2$

Adapted from Stankowski [34].

α, the relation could also be used for less symmetric ligands, at least at low binding densities; see Stankowski [35] for further details.

Cooperative binding presented further mathematical difficulties. Stankowski assumed that the degree of cooperativity would depend on the length of the region of contact between adjacent ligands, that is, on the number of pairwise interactions formed between a ligand and its immediate neighbors. Each of these pairwise interactions was represented by a cooperativity parameter η; $\eta > 1$ would represent positive cooperativity, $\eta = 1$ would indicate no cooperativity (only site exclusion), and $\eta < 1$ would represent negative or anticooperativity. The final result for the binding equation was

$$k[\text{L}] = \eta^{-\delta} B_0 \left(1 + \frac{A}{p}\right)^{\delta} (1 + Ap)^{-\gamma - \delta}$$

(5.28)

where B_0 was defined by

$$B_0 = \frac{<v>}{1-<v>} \left[\frac{1-(1-\lambda)h<v>}{1-h<v>}\right]^{\gamma}$$

(5.29)

and other quantities were defined as follows:

$$\gamma = \frac{\alpha}{h}, \quad \delta = \frac{zh\lambda}{2}, \quad A = \frac{S-1}{S+1}, \quad \text{and} \quad p = \frac{\lambda h<v>}{1-h<v>}$$

(5.30)

Also, the parameter S was defined as

$$S = \sqrt{1 + \frac{4(\eta - 1)\lambda X_2(1 - X_2)}{[1 - (1 - \lambda)X_2]^2}}$$

(5.31)

Qualitatively, the theoretical isotherms for large ligands behave much as has been seen for one-dimensional binding systems, e.g., the McGhee-von Hippel model [1]. In the absence of positive cooperativity, Scatchard-type plots would be concave-upward, just as with the McGhee-von Hippel treatment of noncooperative binding in one-dimensional systems. The larger the ligand, the greater the curvature in the plot would be. The type of lattice (square, hexagonal, honeycomb) apparently does not much affect the curvature of the isotherm in a Scatchard-type plot. The y-axis intercept can be used to obtain the binding constant, much as before, but with a numerical correction for the lattice symmetry. Elongated ligands tend to pack very poorly on a two-dimensional lattice, so that Scatchard-type plots for linear ligands show extreme curvature (see Figure 5.13). For large ligands it would be very difficult to deduce the site size h from the x-axis intercept in a Scatchard-type plot. Unlike the case of linear lattices in which the x-axis intercept is at $1/h$, the intercept for two-dimensional lattices depends on $1/h^2$. Stankowski gave some rough-and-ready means of estimating the site size from the slope of the plot and its projected intercept on the x-axis.

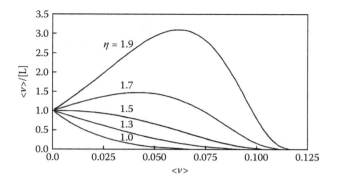

FIGURE 5.13 Curvature in Scatchard plots for two-dimensional lattice binding. Binding isotherms for a linear ligand ($h = 8$) with different values of the cooperativity η; in ascending order of the curves, $\eta = 1.0$, 1.3, 1.5, 1.7, and 1.9.

When positive cooperativity in two dimensions is introduced, the Scatchard-type plot can show a characteristic hump. As with one-dimensional systems, positive cooperativity can compensate (at least partially) for the effects of site exclusion, but this will depend on the shape of the ligand. The plot will show a hump when

$$\eta > 1 + \frac{(1 + \gamma\lambda)}{zh\lambda^2} \tag{5.32}$$

Thus, a lesser degree of cooperativity is needed for linear ligands than for ligands with a more compact shape. This is because, in two dimensions, linear ligands can make more nearest-neighbor contacts than can compactly shaped ligands.

5.2.2 APPLICATION TO MEMBRANE BINDING

Stankowski applied his qualitative model to several different membrane-ligand systems using data drawn from the literature. Melittin, a peptide found in bee venom, binds strongly to lipid membranes; the binding of melittin to vesicles of dimyristoyl phosphatidylcholine (DMPC) was characterized by Vogel [38]. DMPC vesicles undergo a phase transition near room temperature under the conditions used in this study, and the peptide binding is qualitatively different below and above the transition. To interpret his results, Vogel used a simple mass-action model (which of course does not take into account size-exclusion effects), and from this suggested that the number of sites occupied per peptide was significantly greater for the low-temperature phase (59 at 26°C, and 96 at 15°C). These site sizes, however, are inconsistent with other investigations. Stankowski modeled this binding system by assuming that the peptide bound as a large linear ligand [33] and found much smaller values for the site sizes (a value of 21 at 26°C, and 29 at 15°C). This is mainly a result of accounting more accurately for the effects of site exclusion. If these site sizes are divided by a factor of 2, to account for peptide binding only to the outside of the vesicles, then they agree with the results of other investigations.

In another application, Stankowski modeled the binding of the antibiotic polymyxin B to phosphatidic acid bilayers [35]. This system exhibits positive binding cooperativity [39]. Assuming that the lipid lattice was hexagonal, Stankowski was able to fit the experimental binding isotherm with models that used an $h = 7$ hexagon that had a linear protrusion of two units, or a linear ligand shape with $h = 8$, or a compact ligand shape with $h = 10$. Values of the binding constant were similar for all three models (K_0 values from 280 M^{-1} to 530 M^{-1}), as were values for the cooperativity parameter (η values from 1.59 to 1.72). Apparently, the binding in this system is not very sensitive to shape. It is important to note, however, that although the values for η seem to be relatively small, values for η^δ are on the order of 10^3 here, showing the high degree of cooperativity possible when particles have neighbors in two dimensions.

Tamm and Bartoldus [40] have investigated the binding of antibodies to rhodamine B-labeled liposomes, at various concentrations of added labeled lipid (the hapten for the antibody). They showed that, while a Langmuir isotherm equation could be fit to the data at any single concentration of added hapten, when results were compared across different hapten concentrations it appeared that the binding constant and stoichiometry depended on the hapten density. These inconsistencies were resolved when they applied the Stankowski model for disklike ligands [34].

Finally, it is worth noting an alternative to lattice-binding treatments, one that models the adsorbed ligands as a two-dimensional fluid [41]. Now the ligands are not localized to particular lattice points, but instead are permitted to move about freely in two dimensions, just as particles in a gas or liquid would move in three dimensions. Site exclusion and ligand-ligand attractions or repulsions are handled now in much the same way that steric repulsions and noncovalent attractions are treated in the three-dimensional case. Chatelier and Minton [41] have used this approach in modeling the adsorption of globular proteins onto a planar surface, modeling the proteins as either nonassociating or self-associating hard spheres (hard discs, when bound to the surface).

REFERENCES

1. McGhee, J.D. and von Hippel, P.H., Theoretical aspects of DNA-protein interactions: Co-operative and non-co-operative binding of large ligands to a one-dimensional homogeneous lattice, *J. Mol. Biol.* 86, 469, 1974.
2. Kowalczykowski, S.C. et al., Interactions of bacteriophage T4-coded gene 32 protein with nucleic acids. I. Characterization of the binding interactions, *J. Mol. Biol.* 145, 75, 1981.
3. Bujalowski, W., Lohman, T.M., and Anderson, C.F., On the cooperative binding of large ligands to a one-dimensional homogeneous lattice: The generalized three-state lattice model, *Biopolymers* 28, 1637, 1989.
4. McGhee, J.D., Theoretical calculations of the helix-coil transition of DNA in the presence of large, cooperatively binding ligands, *Biopolymers* 15, 1345, 1976.

5. LeBret, M., Numerical solution of McGhee and von Hippel equations for competing ligands, *Biochemistry* 17, 5087, 1978.

6. Schwarz, G. and Stankowski, S., Linear cooperative binding of large ligands involving mutual exclusion of different binding modes, *Biophys. Chem.* 10, 193, 1979.

7. Tsuchiya, T. and Szabo, A., Cooperative binding of *n*-mers with steric hindrance to finite and infinite one-dimensional lattices, *Biopolymers* 21, 979, 1982.

8. Chen, Y.-d., Binding of *n*-mers to one-dimensional lattices with longer than close-contact interactions, *Biophys. Chem.* 27, 59, 1987.

9. Chen, Y.-d., A general secular equation for cooperative binding of *n*-mer ligands to a one-dimensional lattice, *Biopolymers* 30, 1113, 1990.

10. Tsodikov, O.V. et al., Analytic binding isotherms describing competitive interactions of a protein ligand with specific and nonspecific sites on the same DNA oligomer, *Biophys. J.* 81, 1960, 2001.

11. Lifson, S., Partition functions of linear-chain molecules, *J. Chem. Phys.* 40, 3706, 1964.

12. Braunlin, W.H., Strick, T.J., and Record, M.T. Jr., Equilibrium dialysis studies of polyamine binding to DNA, *Biopolymers* 21, 1301, 1982.

13. Latt, S.A. and Sober, H.A., Protein-nucleic acid interactions. II. Oligopeptide-polyribonucleotide binding studies, *Biochemistry* 6, 3293, 1967.

14. Alma, N.C.M. et al., Fluorescence studies of the complex formation between the gene 5 protein of bacteriophage M13 and polynucleotides, *J. Mol. Biol.* 163, 47, 1983.

15. Lerman, L.S., Structural considerations in the interaction of DNA and acridines, *J. Mol. Biol.* 3, 18, 1961.

16. Crothers, D.M., Calculation of binding isotherms for heterogeneous polymers, *Biopolymers* 6, 575, 1968.

17. Bresloff, J.L. and Crothers, D.M., DNA-ethidium reaction kinetics: Demonstration of direct ligand transfer between DNA binding sites, *J. Mol. Biol.* 95, 103, 1975.

18. Bresloff, J.L. and Crothers, D.M., Equilibrium studies of ethidium-polynucleotide interactions, *Biochemistry* 20, 3547, 1981.

19. LePecq, J.-B. and Paoletti, C., A fluorescent complex between ethidium bromide and nucleic acids. Physical-chemical characterization, *J. Mol. Biol.* 27, 87, 1967.

20. Chou, W.Y. et al., The thermodynamics of drug-DNA interactions: Ethidium bromide and propidium iodide, *J. Biomolec. Struct. Dyn.* 5, 345, 1987.

21. Tsai, C.-C., Jain, S.C., and Sobell, H.M., X-ray crystallographic visualization of drug-nucleic acid intercalative binding: Structure of an ethidium-dinucleoside monophosphate crystalline complex, ethidium:5-iodouridylyl(3´-5´)adenosine, *Proc. Natl. Acad. Sci. USA* 72, 628, 1975.

22. Winkle, S.A., Rosenberg, L.S., and Krugh, T.R., On the cooperative and noncooperative binding of ethidium to DNA, *Nucleic Acids Res.* 10, 8211, 1982.

23. Kolb, A. et al., Transcriptional regulation by cAMP and its receptor protein, *Annu. Rev. Biochem.* 62, 749, 1993.

24. Saxe, S.A. and Revzin, A., Cooperative binding to DNA of catabolite activator protein of *Escherichia coli*, *Biochemistry* 18, 255, 1979.

25. Garner, M.M. and Revzin, A., Interaction of catabolite activator protein of *Escherichia coli* with single-stranded deoxyribonucleic acid, *Biochemistry* 20, 306, 1981.

26. Revzin, A., Nonspecific binding and the lactose operon, in *The Biology of Nonspecific DNA-Protein Interactions*, Revzin, A., Ed. Boca Raton, FL: CRC Press, 1990, p. 89.

27. Steitz, T.A., Structural studies of protein-nucleic acid interaction: The sources of sequence-specific binding, *Quart. Rev. Biophys.* 23, 205, 1990.

28. Schultz, S.C., Shields, G.C., and Steitz, T.A., Crystal structure of a CAP-DNA complex: The DNA is bent by 90°, *Science* 253, 1001, 1991.

29. Takahashi, M., Blazy, B., and Baudras, A., An equilibrium study of the cooperative binding of adenosine cyclic 3′,5′-monophosphate and guanosine cyclic 3′,5′-monophosphate to the adenosine cyclic 3′,5′-monophosphate receptor protein from *Escherichia coli, Biochemistry* 19, 5124, 1980.

30. Blazy, B., Culard, F., and Maurizot, J.C., Interaction between the cyclic AMP receptor protein and DNA. Conformational studies, *J. Mol. Biol.* 195, 175, 1987.

31. Takahashi, M. et al., Ligand-modulated binding of a gene regulatory protein to DNA. Quantitative analysis of cyclic-AMP induced binding of CRP from *Escherichia coli* to non-specific and specific DNA targets, *J. Mol. Biol.* 207, 783, 1989.

32. Heyduk, T. and Lee, J.C., Application of fluorescence energy transfer and polarization to monitor *Escherichia coli* cAMP receptor proteins and lac promoter interaction, *Proc. Natl. Acad. Sci. USA* 87, 1744, 1990.

33. Stankowski, S., Large-ligand adsorption to membranes. I. Linear ligands as a limiting case, *Biochim. Biophys. Acta* 735, 341, 1983.

34. Stankowski, S., Large-ligand adsorption to membranes. II. Disk-like ligands and shape-dependence at low saturation, *Biochim. Biophys. Acta* 735, 352, 1983.

35. Stankowski, S., Large-ligand adsorption to membranes. III. Cooperativity and general ligand shapes, *Biochim. Biophys. Acta* 777, 167, 1984.

36. Andrews, F.C., Simple approach to the equilibrium statistical mechanics of the hard sphere fluid, *J. Chem. Phys.* 62, 272, 1975.

37. Andrews, F.C., A simple approach to the equilibrium statistical mechanics of two-dimensional fluids, *J. Chem. Phys.* 64, 1941, 1976.

38. Vogel, H., Incorporation of melittin into phosphatidylcholine bilayers. Study of binding and conformational changes, *FEBS Lett.* 134, 37, 1981.

39. Hartmann, W., Galla, H.-J., and Sackmann, E., Polymyxin binding to charged lipid membranes. An example of cooperative lipid-protein interaction, *Biochim. Biophys. Acta* 510, 124, 1978.

40. Tamm, L.K. and Bartoldus, I., Antibody binding to lipid model membranes. The large-ligand effect, *Biochemistry* 27, 7453, 1988.

41. Chatelier, R.C. and Minton, A.P., Adsorption of globular proteins on locally planar surfaces: Models for the effect of excluded surface area and aggregation of adsorbed protein on adsorption equilibria, *Biophys. J.* 71, 2367, 1996.

6 Choosing a Method and Analyzing the Data

The previous chapters have been concerned with interpreting the results of experimental binding isotherms. How the binding isotherm was determined in the first place was not an issue there, and for the most part this book has ignored experimental details en route to modeling and interpreting the isotherm that was so obtained. This chapter now deals with the choice of experimental method and with general aspects of analysis of the data generated from the selected method.

The discussion here will be concerned with general principles of designing and interpreting a binding study. It is not a survey of experimental methods. Although there will be some discussion of binding kinetics, this chapter will discuss only those major kinetic effects that have an impact on the choice of experimental method, going just far enough to provide some advice for the experimentalist wanting to validate an assay at equilibrium. Surveys of experimental methods may be found in the books by Cantor and Schimmel [1], Campbell and Dwek [2], Connors [3], Hulme [4], Klotz [5], and Winzor and Sawyer [6].

6.1 CONSIDERATIONS WHEN CHOOSING A METHOD

The choice of a proper method for characterizing the binding is an important factor in any binding study. Here are some things one should consider:

1. How selective is the technique? Will it give information about just one kind of binding site, or general information about all sites? For example, methods that separate and measure the free ligand concentration, like equilibrium dialysis, will give information about the average extent of binding, but not about binding to particular sites. Spectroscopic methods, on the other hand, may be sensitive to binding only at one site on the macromolecule and completely insensitive to binding elsewhere.
2. Over what range of concentrations can the technique be used? If the binding is very weak or the method is not very sensitive, relatively high concentrations of ligand and macromolecule must be used. But there may be intrinsic factors, such as the solubility of the ligand or the tendency of a protein to aggregate, that put an upper limit to the range of concentrations of L and M that can be studied. Conversely, with high affinity binding it is necessary to use very low concentrations, and so a method must be chosen that can detect a relatively small change in signal strength.
3. How many separate experiments must be done, and how long will they take? Some methods, like titration calorimetry, can provide a complete binding isotherm in a single run or series of experiments. Other methods, like

equilibrium dialysis, give a single point on the binding isotherm per experiment. Some methods are relatively time consuming (equilibrium sedimentation or dialysis, for example), while others are easy to perform in rapid succession (pH measurements, or radioligand filtration assays, for example).

4. How pure must the system be for useful work? There may be limits on the amount of material available; a receptor may be hard to isolate, for example, or a ligand difficult to label, so that the experimenter might have to settle for some degree of impurity in order to have enough material for the assay. With partially purified systems it is also necessary to take care to avoid competition with endogenous ligands that may have been copurified with the receptor. If the ligands have high selectivity and affinity, it may be possible to use equilibrium dialysis or radioligand binding assays at low ligand concentrations, to reduce the background binding to nonspecific sites of low affinity. With methods that depend on the perturbation of a physical property of either the ligand or the macromolecule, background binding can be more of a problem because substantially higher concentrations of ligand and macromolecule must be used. Table 6.1 summarizes these and other considerations in choosing a suitable experimental method.

6.1.1 Possible Assay Interference from Binding Kinetics

Several of the most popular techniques for assaying macromolecular binding are not themselves true equilibrium techniques, e.g., filtration, electrophoretic methods, and

---

TABLE 6.1
Criteria for Choosing a Binding Technique

Criterion	Comments
Cost	Apparatus initial cost, maintenance, technician salary, supplies and consumables
Convenience	Readily available apparatus vs. specially designed and built equipment, floor space, special facilities like hood, vents, power supply
Time	"Manipulation" time needed for preparing assay (dispensing, mixing, etc.), "intrinsic" time needed for system to equilibrate, overall time needed for assay repetition for statistical validation
Throughput	Possible automation, possibilities for running parallel assays or otherwise multiplexing assays
Type of information gained	Thermodynamic information—which variables? Signal from ligand or from the macromolecule? Does the signal measure stoichiometry?
Range	Upper and lower bounds to K, to amounts and/or concentrations of material used or needed
Noise	Background, controls, filtering, need for repetition and for statistical analysis
Safety	Corrosives, flammable or noxious vapors or solvents, radioactivity, electrical shock, etc.

---

other nonhomogeneous assays that rely on a separation step. With these techniques there is the possibility that the results will not fairly represent the extent of binding. Of course, one wants to make measurements on complexes while they last; if they rapidly dissociate after the separation step, then any delay in applying the chosen method may lead to an underestimate of the amount of complexation and so lead to erroneous estimates of the parameters n and K. Conversely, if equilibration is very slow, as might happen, for example, if binding requires an especially slow conformational change in the macromolecule, then prompt measurements of complexation would also underestimate the extent of complexation, and again lead to an erroneous value for K. Furthermore, some sites may equilibrate more slowly than others, and too-prompt assaying will upset measurements of stoichiometry if all the sites are not equilibrated. On the other hand, if one waits too long, there is the frequent instability of the macromolecule to consider; this can lead to an irreversible denaturation, or to loss of binding activity as either or both the ligand and macromolecule bind nonspecifically to parts of the apparatus. Again, this would lead to erroneous estimates of the binding parameters n and K.

The most important kinetic quantities will be the half-time for association (at a given set of initial concentrations of ligand and macromolecule), and the half-life of the resulting complex. Knowledge of these kinetic parameters will help in choosing an equilibrium binding technique appropriately. Thus, a first step in a binding study should be a preliminary estimate or determination of the rates of association and dissociation.

6.1.1.1 The Rate of Association

The association rate will determine the amount of time a ligand should be incubated with a macromolecule in order to reach equilibrium. A careful experimenter will want to at least make an estimate of this time so as to choose an appropriate incubation period for the assay. For a bimolecular association, the rate will be given by Rate = k_{bimol}[L][M], where k_{bimol} is the bimolecular rate constant. Given an initial concentration of macromolecules (with one binding site per macromolecule, for simplicity), along with knowledge of k_{bimol}, what is the characteristic time for association, and how does it relate to the various experimental methods available?

Neglecting any back-reaction that dissociates complexes, an upper limit to the association rate is set by how fast the ligand and the macromolecule diffuse through solution to encounter one another. For diffusion-controlled association with k_{bimol} equal to about 10^9 $M^{-1}\text{-}s^{-1}$ (probably an overestimate for biochemical systems in which one of the partners is a biopolymer that will diffuse slowly), and both [M] and [L] in the micromolar concentration range, this gives an initial rate of association of about 10^{-3} $M\text{-}s^{-1}$. (This rate won't be sustained, of course, since the available M and L are rapidly depleted.) Using the integrated form of the rate law, one can calculate that within about a tenth of a second the association reaction is essentially (99%) complete under these conditions. So it then seems permissible to assay the system for equilibrium binding after a couple of seconds.

Not all associations proceed under diffusion control, so one might consider a lower value of k_{bimol}, say 10^6 $M^{-1}\text{-}s^{-1}$. For micromolar initial concentrations of M and L this gives an initial rate of association of 10^{-6} M-s^{-1}. The integrated rate law now predicts 99% completion in about 99 seconds. With systems like this it is probably best to wait several minutes before measuring the binding signal. If k_{bimol} is set at 10^3 M^{-1}-s^{-1}, then with micromolar starting concentrations the forward reaction reaches 99% completion after about 10^5 seconds (or roughly 12 days). This would probably be too slow for practical binding assays, due to instability of the materials, especially biopolymers.

6.1.1.2   The Dissociation Rate

The dissociation kinetics are, of course, important when performing assays that separate complex from free ligand, or that involve competition of one ligand with another. The common filtration assay used in pharmacological studies typically can be performed in a matter of seconds. Filtration through a gel matrix may take a few minutes to separate macromolecules from free ligands, while electrophoretic methods may take 10 minutes or more to separate complexes from free ligand. If there is an exponential decay of complexes, corresponding to first-order dissociation kinetics, then as a rule of thumb one should plan to complete the separation part of the assay within a tenth of a half-life of the dissociating complex (93% of the complexes still remain at this point). For a first-order decay process, the relation between the dissociation rate constant k_{diss} and the half-life $t_{1/2}$ is $t_{1/2} = 0.693/k_{diss}$. If k_{diss} equals 10^{-6} s^{-1}, then the complex half-life is about 6.9×10^5 seconds, which means one could allow a leisurely day or two for separation and assay completion. If the dissociation process is speeded up by a factor of a thousand, with k_{diss} equal to 10^{-3} s^{-1}, then the half-life decreases to about 700 seconds; so the assay had better not take more than about a minute to separate ligand from complex. A dissociation with k_{diss} of the order of 1 s^{-1} will be too fast for a manual technique to capture and separate out complexes; these cases call for special instrumentation to follow the reaction.

6.1.1.3   Further Kinetic Considerations

Apart from the rates of association and dissociation, another factor to consider is the rate of conformational changes in the macromolecule. The most extreme change would be denaturation and inactivation of the receptor, but events of lesser magnitude can also have a great effect on the binding. These might include switching of the macromolecule between two or more conformations that have different intrinsic affinities for the ligand, or the dissociation of receptor subunits with consequent changes in ligand affinity.

Clearly, the intrinsic stabilities of the binding partners are important; the rate of inactivation of either partner should be much slower than the assay rate. Since the stability of components may vary considerably with changes in solution conditions, this should be reconsidered every time the pH, salt, temperature, etc. is changed. One happy effect of ligand binding is generally to stabilize macromolecular receptors against denaturation and dissociation; so, even if the isolated receptor is unstable under certain conditions, the complex may persist long enough to permit an assay to succeed.

Kinetic measurements can provide information on ligand binding mechanisms that cannot be obtained from equilibrium measurements. Equilibrium binding assays may not be able to distinguish between anticooperative behavior by a set of identical macromolecules versus the effects of having two kinds of (independent) binding sites present that differ in affinity for ligand. Also, it is extremely difficult to use the shape of an equilibrium binding isotherm for a cooperative, multisite system to determine the mechanism behind the cooperativity, since different mechanisms can lead to what is apparently the same binding curve. (It is well known that this difficulty arises in the binding of so-called large ligands to linear biopolymers such as DNA or RNA; there can be similar difficulties in resolving mechanisms for large enzymes binding substrates or other small molecules.) Kinetic studies combined with careful purification protocols may be able to resolve these ambiguities by, e.g., distinguishing different kinetic phases (in association or dissociation) with different concentration dependencies. Kinetic studies may also be able to detect and characterize conformational changes during and subsequent to binding. Order-of-addition experiments can help resolve questions about the order of binding of multiple ligands. Also, kinetic studies can (under suitable conditions) detect and resolve rapid nonspecific associations from slower, more specific binding of a ligand.

6.1.2 DIRECT VERSUS INDIRECT METHODS OF MEASURING THE EQUILIBRIUM

To measure the extent of binding, one needs to be able to determine quantitative changes in the amount of one or another of the system's components. It might be possible to do this by following that component's thermodynamic activity directly (e.g., by use of a pH meter when following the binding of protons), or it might be possible to track changes in some physical property of that component (its optical absorbance or fluorescence, for example). It would be most convenient if this property were to change in exact proportion to the extent of binding, and if it could be measured without separating components of the system. With such a *homogeneous assay* one could first determine the proportionality constant joining the component's property to its concentration, then simply follow the property's change while titrating the system. In view of the multiple and quite disparate physical properties of different types of ligands and macromolecules, one can hope that there would be some such physical property that serve—and often enough, there is. In other cases, however, a homogeneous assay cannot be devised; this situation then calls for a *heterogeneous assay*, in which components are separated and then measured.

Measurements of changes in a true thermodynamic property, particularly one that reflects the free ligand concentration or activity, give direct access to the binding isotherm. One such method is the use of a reversible electrode (for measuring the chemical potential of an electrochemically active species; e.g., a pH meter for measuring the activity of H^+); others include phase partitioning and equilibrium sedimentation. Another thermodynamic method, with an extensive history of applications, is equilibrium dialysis; results from this very basic method are often used as the standard by which to judge results obtained through other techniques. Finally, there is the standard thermodynamic method of calorimetry, the detection of heat absorption or emission by the system. These methods, however, are often slow or clumsy, or consume much precious material. It is often much more convenient to try to detect and follow a physical change in either ligand or macromolecule as a function of binding.

Physical changes might include a gross conformational change in the macromolecule, or a local shift in electron density about a chromophore in either the ligand or the macromolecule. Often, the physical changes can be detected by spectroscopy of one type or another. There might be a change in UV absorbance of a group upon forming a complex; there might be a change in the circular dichroism or the fluorescence of one species or another, or changes in the NMR or EPR spectrum. Other methods might detect differences in dynamic behavior (diffusion, sedimentation, or electrophoretic mobility).

Fortunately, the signal derived from the physical change is often linearly proportional to the extent of binding, and so it is relatively straightforward to obtain the isotherm from such measurements. It may be necessary to make a preliminary set of measurements, to determine the constant of proportionality, before launching into the ultimate experiment, the titration itself (the determination of n and K). For example, one might need to measure how much the absorbance of a protein will change in going from completely "free" to fully saturated with ligand, or to measure the extent of quenching of its fluorescence at saturation, etc.

Physical changes in properties of the ligand or receptor are really only *indirect* measures of the extent of binding, however, and as such they may fail to reveal the true extent of binding. Suppose that with a certain multisite receptor the first ligand to bind shows a characteristic fluorescence increase. But conceivably, a *second* ligand binding to that same receptor might not affect the overall fluorescence in any noticeable way, due to the spectroscopic properties of the ligand, the nature of the (second) binding site, changes in receptor conformation, etc. This thermodynamic event (the binding of the second ligand) has occurred without a detectable spectroscopic signal to reflect the change in state of the system. Furthermore, binding events with multisite receptors may occur at quite different rates. Then, depending on when the binding is assayed and by which signal, the values of the binding parameters so derived might be quite disparate. In general, determinations of n and K from signals deriving from physical changes in ligand or macromolecule should be treated with skepticism until verified by other methods.

Klotz prepared a useful chart that summarized then-current methods for studying ligand binding by macromolecules [5]. A modified and updated version of his table is given here (Table 6.2).

6.2 DESIGNING THE EXPERIMENT

When starting a binding study an immediate question is, what concentration of ligand and macromolecule should one use? Beyond the obvious condition that there must be present enough signal-generating species to enable detection of the binding, an important question is whether or not it is possible to make measurements across the greater part of the isotherm. Can the experiment be set up (by choosing the proper concentrations of ligand and macromolecule) so as to measure binding below 20% saturation, or up to (and perhaps beyond) 80% saturation? Will this be enough to yield both the binding stoichiometry and the binding affinity?

Also, will there be any interference from nonideal thermodynamic behavior in the system? Can one still use simple concentrations to calculate K and to model isotherms,

TABLE 6.2
Macromolecular Binding Methods

Determination of Concentration of Free Ligand	Perturbation of Properties of Bound Ligand	Perturbation of Properties of Binding Macromolecule
Physical separation of ligand and complex Dialysis Filtration Ultrafiltration Centrifugation Phase partitioning	Spectroscopic Optical absorbance Optical rotation (ORD, CD) Intrinsic fluorescence Fluorescence polarization NMR ESR	Spectroscopic Optical absorbance Optical rotation Surface plasmon resonance (SPR) Fluorescence NMR
Electrochemistry Selective electrodes Perm-selective membranes Polarography Conductivity		Thermodynamic Osmotic pressure Light scattering Equilibrium sedimentation Calorimetry
Biological assay		Dynamic Sedimentation velocity Electrophoresis Light scattering Viscosity Nitrocellulose filter binding
		X-ray diffraction
		Biological assay

Adapted from Klotz [5].

or are thermodynamic activity corrections needed? In most cases one should try to work with dilute solutions, to avoid having to correct for nonideal behavior. This may not be desirable, however, when trying to mimic conditions in vivo, in which the effects of excluded volume and of ionic interactions may become quite pronounced. In living systems the solutions are highly nonideal, and this can grossly affect the position of a binding equilibrium. To interpret binding results under such conditions it may become necessary to apply explicit activity corrections.

6.2.1 NONIDEAL BEHAVIOR: SALT AND CROWDING EFFECTS

Macromolecules like proteins or nucleic acids have a complicated array of charges, with dielectric discontinuities and local variations in polarizability. The electrostatic interactions for these macromolecules are a major feature in their structure and function, but subtleties in these interactions are not easily captured by simple electrostatic models and so may be overlooked or ignored in elementary calculations of activity corrections. Furthermore, whether measured or calculated, activity coefficients alone are not sufficient to explain patterns or specific details of stability, specificity, and kinetic effects for these macromolecules. For a deeper understanding of electrostatic effects in macromolecular binding, the usual approach is to turn to direct electrostatic calculations, using, for example, the Poisson-Boltzmann equation [7,8]. These electrostatic calculations are not directed specifically toward predicting activity coefficients nor at generating binding isotherms, but instead are aimed at finding the overall change in the binding free energy of the system as a function of pH, salt, dielectric constant, etc. As such, these computer model calculations lie beyond the scope of this book, which is concerned with binding isotherms and their interpretation.

Salt effects in the binding of ligands to charged linear polymers are often quite striking in their magnitude and in their dependence on ionic and polyionic parameters [9–11]. The extent of binding of proteins, peptides, drugs, polyamines, and other oligoions to linear polyelectrolytes like DNA typically decreases quite drastically as the concentration of simple salt MX increases. Small changes in the salt concentration, even with dilute salt solutions, can easily change the apparent binding by an order of magnitude or more. The magnitude of these effects indicates that a large number of counterions are released (presumably these come mainly from the polyelectrolyte). The contribution of counterion release to the salt dependence of binding in these systems has been dubbed by Record and coworkers as the "polyelectrolyte effect," by analogy to the well-known hydrophobic effect [12].

Additionally, electrostatic effects play an important role in solvation free energies and in salt effects on the titration of ionizable moieties in macromolecules in general. Simple formulas for activity corrections generally require that the solution be dilute. The classical Debye-Hückel theory of electrolyte solutions gives simple formulas for ionic activities, but these are accurate only for very dilute solutions of simple electrolytes; see Chapter 23 of Lewis and Randall's *Thermodynamics* [13], or Chapter 12 of Butler [14], for an introduction to the basic concepts and formulae. For more concentrated solutions, for mixtures of electrolytes, and for multivalent ions, the size of the ions and their hydration must be taken into account, and the calculations can become rather involved [15,16].

Volume occupancy ("excluded volume") by macromolecules is now recognized as a major factor in macromolecular equilibria in vivo. For typical bacterial or eukaryotic cells, the cytoplasm contains between 17% and 26% protein by weight [17], and though the molarity of individual protein species may be 10^{-3} or lower, the protein molecules overall will occupy an appreciable fraction of the volume; hence, these solutions are *crowded*, though dilute. This *macromolecular crowding* can drive aggregation equilibria toward more compact structures, disfavoring elongated or extended structures. Other effects include the increased stability and persistence of activity of a variety of enzymes under otherwise inhibitory conditions of salt, temperature, or pH; alterations in enzyme specificity; decreases in the solubility of nucleic acids in the presence of added "inert" polymer such as polyethylene glycol; and alterations in the degree of association of proteins such as RNA polymerase with nucleic acids [18–20]. Relatively simple formulas are available for activity corrections due to excluded volume effects [20].

For *small* ligands that are *not* highly charged and that have moderate to strong binding affinity, activity corrections should probably not be a major concern. An empirical check on the need, if any, for activity correction is to perform the binding studies at two very different concentrations of macromolecule when titrating with a small ligand, or with two different ligand concentrations when titrating with a receptor. If the binding constant varies with concentration, then appropriate activity corrections must be sought.

6.2.2 CHOOSING WORKING CONCENTRATIONS IN RELATION TO K

In an ideal experiment the binding data should extend from negligible binding to nearly complete binding saturation, but in practice it is often difficult to collect binding data at these two extremes. At low degrees of binding it may not be possible to detect complexation because the detectors are not sensitive enough. At the other extreme, particularly for weakly binding systems, binding saturation may be approached only with physically unrealizable amounts of ligand (amounts exceeding the solubility limit of the ligand, for example). Furthermore, in this concentration regime, those impurities that were negligible at low concentrations may now become a major interference. Finally, corrections for thermodynamic nonideality (activity coefficients) become increasingly important as the concentration rises; this is one of the major advantages to working in dilute solution in the first place, since activity corrections won't usually be necessary then.

To follow a signal from one component or another across the titration, there should, of course, be enough of that component present so that, first, the signal strength from it will be detectable, and second, there are reliable changes in signal strength as the degree of binding saturation rises. For a signal whose strength is proportional to the amount of material converted to complexes, the signal intensity is $S = \alpha \times Y \times C$, where C is the concentration of receptors, Y is the degree of binding saturation, and α is a proportionality constant that will differ from technique to technique. For example, NMR is a relatively insensitive technique and will require substantially higher concentrations of material (typically in the millimolar range) than typical radiotracer methods (whose high sensitivity permits their use down to the nanomolar range).

Statistical analyses of the effects of experimental design and of errors in titrations suggest that the most reliable estimate of the affinity is obtained when there are roughly equal amounts of "bound" and "free" for the species that generates the signal, whether it is the ligand or the macromolecule. The relative error in the dissociation constant K_{diss} grows as the ratio of "free" to "bound" becomes very large or very small, that is, in the limit of zero binding or of binding saturation. The statistical analyses indicate that the most accurate values of K_{diss} are obtained when Y lies between 0.2 and 0.8 [21–25]. On the other hand, for a determination of the stoichiometry alone it is better to work near $Y = 1$, say, at 80% saturation and above.

For a simple Langmuir single-site isotherm, a spread of free (not total) ligand concentrations from 0.2 K_{diss} to 5 K_{diss} will correspond to a range in Y from 1/6 up to 5/6.

This isn't quite enough to determine the stoichiometry comfortably, so the experimental design should probably be extended for [L] down to 1/10 of K_{diss}, and up to 10 K_{diss}. This hundredfold range in [L] covers both the region where one can expect to get the most reliable estimates of K_{diss} and the region near saturation, from which one can hope to get good estimates of the binding stoichiometry. For a 1:1 binding stoichiometry, this will provide a range in Y from about 0.1 to 0.9. Higher concentrations than 10 K_{diss} will not yield much more information about saturation; measurements at concentrations lower than 1/10 of K_{diss} will probably be difficult to make.

What spacing of the concentrations should be used? Linear spacing consumes much material and requires too many separate experiments (especially if they are done in triplicate, for confidence in the measurement) to cover the hundredfold range of concentrations just suggested. Geometric spacing is often used, e.g., simply doubling the concentration from point to point; another reasonable choice is to use a multiplier of 1.6 instead of 2, to give a denser array of data points. Cleland [26] has suggested collecting data at points evenly separated on a reciprocal plot, especially for concentrations in the region around K_{diss}; his suggestion was made in the context of enzyme kinetic experiments, in which the data might be analyzed by a Lineweaver-Burk (double-reciprocal) plot. This provides a reasonable compromise between density of data points and effort in collecting sufficient data. For a plot of Y as a function of the logarithm of the free ligand concentration (the format suggested by Klotz [27]), a simple geometric progression in [L] will not give evenly spaced points. Instead, one might consider choosing [L] such that $[L]/K_{diss}$ equals 0.100, 0.158, 0.251, 0.398, 0.631, 1.000, 1.58, 2.52, 3.98, 6.31, and 10.00. This spans two orders of magnitude in ligand concentration, with the points evenly spaced on the log [L] axis.

It is often convenient to prepare a set of concentrations by serial dilution, for example, using a robotic sample preparation system. With robotic systems a standard assay design is based on the 96-well microtiter plate. This is a multi-well sample incubation plate with wells arrayed as eight rows of 12 wells each (the standard 8×12 format). This permits side-by-side comparison of eight different ligands for a receptor, at 12 concentrations each. Ordinarily, one well in each sample row is reserved for the "blank," with no ligand added, which leaves 11 wells and 11 different concentrations per ligand. A "half-log" dilution series, with a stepwise dilution factor of $3.16 = 10^{1/2}$, will give evenly spaced data in a semilog plot of Y as a function of *total* ligand concentration (*not* free ligand concentration; see Figure 6.1). Serial dilution could thus be used to prepare samples with total ligand concentrations at 100 μM, 31.62 μM, 10 μM, 3.162 μM, etc.

Finally, all the above recommendations for choices of concentrations have been couched in terms of the dissociation constant, K_{diss}. It is now apparent that one needs at least a preliminary estimate of the value of K_{diss}. If there are already data at hand on K_{diss} for related ligands, then perhaps an average of these can serve as a preliminary estimate.

Sometimes kinetic data on rates of dissociation and association are available. The ratio of the rate parameters for dissociation to association can then give an estimate of K_{diss}. If there are no data on the rate of association, then it is still possible to estimate a preliminary value for the association rate parameter. For example, one

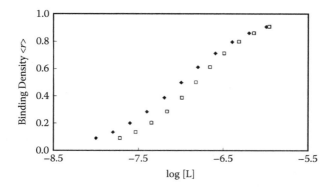

FIGURE 6.1 Comparison of binding density as a function of free versus total ligand concentrations for a single-site model with $k = 1 \times 10^7$ M⁻¹. Free ligand concentrations were chosen to achieve even spacing in a semilog plot ($k \times$ [L] equals 0.100, 0.158, 0.251, 0.398, 0.631, 1.000, 1.58, 2.52, 3.98, 6.31, and 10.00). (◆), free ligand concentration; (□), total ligand concentration.

might assume that association is diffusion controlled, and then standard formulas could be used to estimate the association rate parameter. To complement this, it would also be necessary to make some trial dissociation kinetic experiments, perhaps measuring the half-life of the complex, to get an estimate of an effective dissociation rate constant; then one could proceed to estimate K_{diss}.

6.2.3 SOME GENERAL PRECAUTIONS TO CONSIDER

All experimenters should be aware of the effects of experimental design on the experimental error and its propagation. For example, the kinetic effects of ligand release become highly important in the error analysis for binding assays done using rapid filtration. Furthermore, in radioligand binding assays done by rapid filtration, the error analysis of an experiment using only "hot" ligand will be rather different from that for the case in which "hot" is diluted with "cold" ligand. The purity of the "hot" ligand and its specific activity become critical when analyzing the "hot only" type of experiment, but these are less important when diluting the hot material with cold ligand, so that the bulk of the binding is due to the cold ligand. Of course, in both types of experiments the effect of a possibly incomplete separation of free ligand from bound should be considered.

Next, it is generally desirable to do multiple determinations of the binding at each concentration, at least in duplicate and preferably in triplicate. This will give a rough estimate of the variance in the measurements, which will be useful when performing statistical curve fitting. Furthermore, *all* the original measurements should be submitted to statistical analysis, without preaveraging the data. If there are "outliers," seemingly aberrant points, these should not be discarded or ignored without objective justification, and there are statistical tests to guide this choice. Outliers may in fact indicate the failure of the model to encompass some real behavior on the part of the system. Perhaps the type of signal used is subject to wide fluctuations, so that one should expect a large error; or perhaps there is some sort of large and fundamental change occurring in the system that was not in the model (e.g., some sort of cooperative behavior producing a much larger than expected response).

Finally, it is always a good idea to characterize a binding system by at least two independent experimental methods, to guard against possible inherent biases in one technique or the other. Filtration methods, for example, may be susceptible to artifacts from washing away weakly bound material, and calorimetric methods may need correction for heats of ionization of the buffering agent, etc. Additionally, one method may be more sensitive in, say, the region of low binding density or at low ligand concentrations, while the other method is better suited to high binding densities or concentrations. For example, one could first carry out a titration by holding [M] fixed and varying [L], perhaps following a spectrophotometric signal from the macromolecule. Then this could be supplemented with experiments in which [L] is fixed, and [M] is varied, as, for example, using surface plasmon resonance to follow the binding. The first method is probably better suited to measurements at low concentrations of macromolecule, and the other to high concentrations; furthermore, the two methods are based on quite different physical properties of the system. If the independent methods agree on the binding behavior (same affinity, stoichiometry, cooperativity, etc.), then one can have considerable confidence in having arrived at a reliable isotherm for the system.

6.3 MODEL-FREE ANALYSES OF BINDING SIGNALS

Many experimental methods (e.g., spectroscopic and calorimetric techniques, or measurements of the colligative properties of a solution) will not give a direct measurement of the amount of ligand or macromolecule that is free or bound; these quantities must be determined indirectly by calculation after the measurement. However, these assays (especially spectroscopic measurements) are often simple to perform and usually take relatively little time, and so they are widely used. For thermodynamics the natural independent variable is the free ligand concentration, but with these methods the independent variables now are the *total*, not the free, concentrations of ligand and macromolecule. It is only after the experimental measurement that it is possible to calculate the free concentrations, by subtracting the amount that is apparently bound (as measured in the experiment) from the total amount of the species analyzed (which, presumably, is known beforehand with suitable accuracy). There can, however, be appreciable error in determining the amount bound, particularly at very low or very high levels of binding saturation, so that the free amount may be rather uncertain, too.

With spectroscopy and with other methods in the same class, what is captured is a signal whose change is (presumably) proportional to the degree of binding. Rather than fitting an isotherm equation directly, using free ligand concentrations, one must instead express the observed signal (absorbance, fluorescence, etc.) as a function of the total concentration of the species present and under observation. For these methods the observed signal should be treated as the dependent variable, which is a function of the total concentration of ligand and/or macromolecule (depending on how the titration has been set up). The law of mass action (or some variation of it, if the system has multiple classes of sites, with site-site interaction, etc.), together with mass conservation equations for the ligand and its binding partner(s), can then be used to express the functional dependency.

6.3.1 Assumptions and Notation

Several research groups have independently proposed general methods of analyzing a binding signal when that signal is a linear function of species' concentrations [28–34]. In general, these methods are based on paired measurements taken from at least two different titrations. The pairing of observations relies on having chosen operating concentrations so that the greater part of the isotherm is represented in each titration. These methods link the independent variables of total ligand mass and total macromolecule mass at a given level of signal, and they are in general independent of any particular binding model. Apart from the linearity assumption, and the neglect of any aggregation of ligand or macromolecule, these are in fact model-free methods of analysis that yield a binding isotherm; that is, they can give the degree of saturation as a function of free ligand concentration. This isotherm can then be analyzed in terms of particular binding models to extract stoichiometry, affinity, and cooperativity.

To develop and explicate these concepts, this section will detail the method described by Lohman and Bujalowski [31,33], but with a few changes in notation. A subscript f will be introduced here, matching the subscript t, to avoid confusion between free and total amounts of ligand or macromolecule. Thus, the total ligand concentration will be written as $[L]_t$ and the free macromolecule concentration as $[M]_f$. Also, the macromolecule will be permitted to accept up to n ligands at saturation. For a macromolecule that has i ligands bound ($i < n$), the ligands could be distributed in several different ways over the sites, and depending on precisely which sites were occupied there could be a slightly different signal. An index α will be used for these different kinds or subclasses of complexes that have i ligands bound. The binding density for one such subclass would be written as

$$r_{i,\alpha} = \frac{[L]_{i,\alpha}}{[M]_t} = \frac{i\,[M]_{i,\alpha}}{[M]_t} \tag{6.1}$$

Here the concentration of macromolecule in such complexes has been written as $[M]_{i,\alpha}$ and the concentration of bound ligands within such a subclass as $[L]_{i,\alpha}$. The brackets around the binding density $r_{i,\alpha}$ have been dropped here because the average is not over different kinds of sites, but only over a single kind or subclass of complex.

The concentration of free ligand will determine the distribution of bound ligand and hence the overall binding density. Suppose that there are two different samples containing different total concentrations of macromolecule $[M]_t$. Suppose also that the *total* ligand concentration for each sample has been chosen or manipulated such that they both have the same *free* ligand concentration $[L]_f$. Then the overall binding density $< r >$ in both samples *must also be the same*. This is a key point, the equality of binding density in the two samples, and it is at the base of all the following analyses.

One could go on to set up a third, fourth, etc. sample, each differing in total ligand and total macromolecule concentrations but all with the same total free ligand concentration and the same overall binding density. The general connection between the binding density and the free ligand concentrations in these samples is through conservation of mass of the ligand, from which comes

$$[L]_t = <r>[M]_t + [L]_f \qquad (6.2)$$

If it is possible to find a set of concentration pairs $[L]_t$ and $[M]_t$ for which $[L]_f$ and $<r>$ are constant, then $[L]_f$ and $<r>$ can be determined from the slope and intercept of a plot of $[L]_t$ versus $[M]_t$. This gives a single point on a binding isotherm. Note that this is independent of the particular signal used for detecting the binding, and that no microscopic binding model has been used, only the conservation of mass. Also, the reliability of the pair of values of $[L]_f$ and $<r>$ should increase with more measurements of different concentration pairs $[L]_t$ and $[M]_t$ corresponding to the chosen free ligand concentration and binding density.

A determination of the overall shape of the binding isotherm would need many pairs of values of $[L]_f$ and $<r>$, so that the process would have to be repeated many times. Eventually, however, a full plot of $<r>$ as a function of $[L]_f$ could be obtained, which could be evaluated in terms of microscopic binding models for affinity, stoichiometry, cooperativity, etc. Usually, one can measure and manipulate $[L]_t$ and $[M]_t$ fairly readily, so in principle the approach seems quite feasible. The obvious next question is, how does one find that set of $[L]_t$ and $[M]_t$ values in order to generate the pairs of values of $[L]_f$ and $<r>$, and so construct the isotherm plot? This will depend on whether the experimental method uses a signal from the macromolecule or from the ligand.

6.3.2 SIGNAL FROM THE LIGAND

When the signal derives from the ligand, one should plan to start the titration with a solution containing only ligand at a total concentration $[L]_t$, and to add macromolecules to this in defined increments. The signal observed here might be a fluorescence intensity change of the ligand, a change in its UV absorbance, etc. Suppose that, whatever the method, it yields a signal S that is proportional to the individual concentrations of all the ligand species participating in the binding equilibrium. (This treatment will ignore contributions from unliganded macromolecules; the correction is easy enough to make.) The signal S is given by

$$S = s_f[L]_f + \sum_i \sum_\alpha s_{i,\alpha} [L]_{i,\alpha} \qquad (6.3)$$

where s_f is the signal intensity from unbound (free) ligand, $s_{i,\alpha}$ is the signal intensity from the α-th type of complex with i ligands bound, and $[L]_{i,\alpha}$ is the concentration of bound ligands in those complexes. The double sum extends over all the different bound forms of the ligands. However, no allowance is made here for ligand aggregation in solution or for other possible perturbations of the concentrations of free and bound ligand. The different signal intensities might come from different conformational states of the macromolecule, or from different kinds of binding sites on a macromolecule, etc.

Mass conservation for the ligand requires

$$[L]_t = [L]_f + \sum_i \sum_\alpha [L]_{i,\alpha} \tag{6.4}$$

Use of this relation, together with that relating $[L]_{i,\alpha}$ to the total concentration of macromolecule, gives the connection between the observed signal and the contributions from all species present:

$$S = s_f[L]_t + [M]_t \sum_i \sum_\alpha (s_{i,\alpha} - s_f) \times r_{i,\alpha} \tag{6.5}$$

which can be recast in terms of a relative change in signal as

$$\Delta S \times \frac{[L]_t}{[M]_t} = \sum_i \sum_\alpha \Delta s_{i,\alpha} r_{i,\alpha} \tag{6.6}$$

where

$$\Delta S = \frac{S - s_f[L]_t}{s_f[L]_t} \tag{6.7}$$

and

$$\Delta s_{i,\alpha} = \frac{s_{i,\alpha} - s_f}{s_f} \tag{6.8}$$

Lohman and Bujalowski refer to the quantity ΔS ($[L]_t/[M]_t$) as the ligand binding density function or LBDF. $\Delta s_{i,\alpha}$ is the signal change per bound ligand in the complex containing i ligands of the α-th type, while ΔS is the observed fractional change in signal from all ligands, with the solution containing a total ligand concentration $[L]_t$ and total macromolecule concentration $[M]_t$. The quantity s_f $[L]_t$ is the signal observed at the start of the titration, before addition of macromolecule. Notice that the quantity $\Delta s_{i,\alpha}$ is an intrinsic property of the type of complex formed; it should be constant for a particular binding state $\{i,\alpha\}$ under a given set of experimental conditions. The corresponding sum $\Sigma\Sigma\ \Delta s_{i,\alpha}\ r_{i,\alpha}$ will be a constant for the specified distribution of binding densities $\Sigma\Sigma\ r_{i,\alpha}$. For a given value of ΔS ($[L]_t/[M]_t$), at equilibrium the values of $\Sigma\Sigma\ r_{i,\alpha}$ and of $[L]_f$ are constant and independent of the total macromolecule concentration. This means that if one has two or more independent titrations, done at different total ligand concentrations, one can obtain model-independent values (estimates, really) of $\Sigma\Sigma\ r_{i,\alpha}$ and $[L]_f$ from plots of ΔS ($[L]_t/[M]_t$) against $[M]_t$, so long as the pH, temperature, pressure, etc. are the same for the titrations.

To determine $\Sigma\Sigma\ r_{i,\alpha}$ and $[L]_f$, one plots the two (or more) titrations on a single graph with ΔS ($[L]_t/[M]_t$) on the ordinate (y-axis) and $[M]_t$ on the abscissa (x-axis) (see Figure 6.2A). Individual smooth curves can then be fit through each set of data, to aid in interpolation; these could be simple polynomial fits, and at this point it is not necessary to make any suppositions about a binding model for the isotherm. It should be clear that a horizontal line connecting two titration curves would correspond to a single value of the LBDF, one that is shared by both titrations. Also, the x-axis values of the points at which this horizontal line intersects the two titration curves would define two values of $[M]_t$ at which the binding density was the same. This gives two pairs of values, ($[L]_{t1}$, $[M]_{t1}$) and ($[L]_{t2}$, $[M]_{t2}$), for a secondary plot of $[L]_t$ as a function of $[M]_t$. The slope of this secondary plot (Figure 6.2B) equals the overall binding density $< r >$, and the intercept on the ordinate equals the free ligand concentration $[L]_f$.

If there were more pairs of values of ($[L]_f$, $< r >$), one could then go on to construct a tertiary plot of $< r >$ as a function of $[L]_f$. This tertiary plot is, at last, the binding isotherm itself (see Figure 6.2C). The data could also be plotted in another format, as a Scatchard plot for example, if so desired. To obtain more points on the isotherm, the analytic process would have to be repeated on the primary plot, choosing, of course, a different value of the LBDF.

More accuracy in those isotherm points (that is, more accurate estimates of $[L]_f$ and $< r >$ from the secondary plot) would require a third, fourth, etc. titration curve in the primary plot. This would allow better estimates of the slope and intercept of the secondary plot. By using fitted smooth curves on the primary plot, one can interpolate values and avoid having to collect data at precisely matching values of the LBDF and of $[M]_t$.

The different titration experiments may use quite different values of total ligand concentration and hence arrive at quite different total macromolecule concentrations. As a practical matter, then, it is convenient to compress the values of $[M]_t$ by using a logarithmic scale.

6.3.3 Signal from the Macromolecule

When the signal derives from the macromolecule (fluorescence, absorbance, etc.), it is common practice to hold constant the concentration of the macromolecule and to vary the total ligand concentration in the titration. For model-free determination of the isotherm, Lohman and Bujalowski again use a binding density function, but here it is the macromolecule binding density function or MBDF.

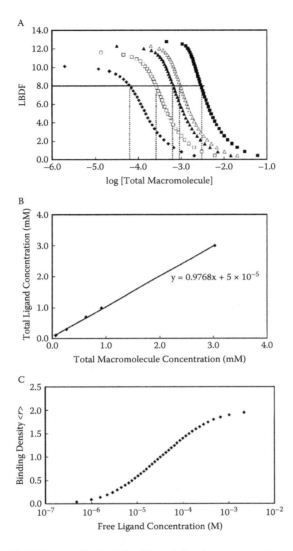

FIGURE 6.2 Model-free analysis of families of titration curves when an absorbance signal derives from the ligand. A. Hypothetical LBDF for two unequal, independent binding sites per macromolecule, with $k_1 = 1.0 \times 10^5$ M^{-1} and $k_2 = 1.0 \times 10^4$ M^{-1}, as a function of total macromolecule concentration $[M]_t$. Horizontal line connects a set of values, $([L]_{t1}, [M]_{t1})$, $([L]_{t2}, [M]_{t2})$, etc., at which the binding density is the same. For purposes of modeling, the assumed extinction coefficients were ε (free ligand) $= 1.0 \times 10^3$, ε (bound ligand, site 1) $= 1.0 \times 10^4$, and ε (bound ligand, site 2) $= 5.0 \times 10^3$. Total ligand concentrations (molar) were $[L]_{t1} = 1.0 \times 10^{-4}$ (◆), $[L]_{t2} = 3.0 \times 10^{-4}$ (□), $[L]_{t3} = 7.0 \times 10^{-4}$ (▲), $[L]_{t4} = 1.0 \times 10^{-3}$ (△), and $[L]_{t5} = 3.0 \times 10^{-3}$ (■). B. Schematic secondary plot of $[L]_t$ as a function of $[M]_t$, for the horizontal line of panel A. The slope equals the overall binding density $< r >$, and the intercept on the ordinate equals the free ligand concentration $[L]_f$. C. Binding isotherm, $< r >$ as a function of $[L]$, that would be obtained by extension of plots of the type shown in panel B.

The observed signal S (from the macromolecule) is given by

$$S = [M]_t \left(s_f + \sum_i \sum_\alpha^n \left(s_{i,\alpha} - s_f \right) \frac{r_{i,\alpha}}{i} \right) \tag{6.9}$$

which can be recast in terms of ΔS, a relative change in observed signal, as

$$\Delta S = \frac{S - s_f[M]_t}{s_f[M]_t} = \sum_i \sum_\alpha \Delta s_{i,\alpha} \left(\frac{r_{i,\alpha}}{i} \right) \tag{6.10}$$

where

$$\Delta s_{i,\alpha} = \frac{s_{i,\alpha} - s_f}{s_f} \tag{6.11}$$

$\Delta s_{i,\alpha}/i$ is the average signal change per bound ligand in the complex containing i ligands of the α-th type. The quantity $s_f\,[M]_t$ is the signal observed at the start of the titration, before addition of ligand. The quantity ΔS is the MBDF. Notice that ΔS is a function only of the intrinsic properties of the macromolecule (through the $s_{i,\alpha}/i$) and of the binding densities $r_{i,\alpha}$. This is analogous to the dependence of the LBDF on the binding density; it implies that, for two (or more) titrations at different total macromolecule concentrations, the value of $[L]_f$ is the same when the MBDF is the same for the titrations.

The procedure for obtaining pairs of values ($[L]_f$, $<r>$) is much the same as for the situation in which the signal is from the ligand: carry out at least two titrations at different total concentrations of macromolecule, then plot the MBDF as a function of total ligand concentration, and perhaps fit individual smooth curves through the titration data for use in interpolation. Draw a horizontal line to connect the curves, to obtain pairs of values of ($[L]_{t1}$, $[M]_{t1}$) and ($[L]_{t2}$, $[M]_{t2}$), at a common value of the free ligand concentration and the binding density. Use these values in a secondary plot of $[L]_t$ as a function of $[M]_t$, and from the slope and intercept of this secondary plot, deduce values for $<r>$ and $[L]_f$. Repetition of the process will eventually permit construction of a tertiary plot of $<r>$ as a function of $[L]_f$.

The LBDF or MBDF approach yields a thermodynamically rigorous binding isotherm. It is possible to extract the values of ($[L]_f$, $<r>$) from the data without making any assumptions about the underlying binding mechanism, apart from the assumption that the ligand does not aggregate appreciably or is otherwise perturbed. The general method makes no assumptions about the numbers of sites on the macromolecule, the strength of their interactions with the ligand, or their interactions with each other (cooperativity). The binding isotherm derived this way can then be analyzed in terms of the microscopic models discussed in Chapters 2 through 5, using nonlinear curve-fitting methods.

A point worth noting with this general approach is that, for either type of titration, one does not need to evaluate the signal value at binding saturation. This quantity is often difficult if not impossible to obtain, due to the high concentrations that may be needed to achieve saturation or to difficulties in measuring signals near binding saturation, and it is often found only by extrapolation, with consequent errors in its determination.

Complicated macromolecular systems with multiple binding states can easily have a plethora of intrinsic signal parameters. For example, the DnaB protein of *E. coli* has six nucleotide binding sites, and a simple hexagon model of this binding system would generate at least 10 molecular quenching constants for fluorescence titrations [35]. With this many parameters it is all but impossible to determine unique individual values for the signal parameters. Instead, a practical approach would be to use an approximate empirical function to relate the average observed signal level to the sum of the binding densities $\Sigma\Sigma\, r_{i,\alpha}$. For simplicity, this function could be a polynomial in $\Sigma\Sigma\, r_{i,\alpha}$. In the case of the nucleotide/DnaB protein system, Bujalowski and Klonowska [35] expressed the observed fluorescence quenching Q_{obsd} as a third-degree polynomial in the binding densities; this function was used to generate theoretical isotherms for comparison to the experimentally determined isotherm, and so to extract the binding constants and binding cooperativities.

Finally, the model-free analysis of ligand binding has been extended to competitive binding systems [36]. This is especially useful for systems in which the binding of one of the competing species does not generate an appreciable signal, while the binding of a second ligand (a reference system, which does generate an observable signal) is monitored.

6.4 STATISTICAL ANALYSIS OF BINDING DATA

The standard for publication of a binding study now includes a statistical analysis by computer of one's binding data. Modern software packages for statistical analysis are widely available; even common spreadsheet programs have a variety of statistical tools, and free supplements ("add-ins") to these spreadsheet programs have been developed (e.g., Raguin et al. [37]). These programs are generally easy to use and do not require much statistical sophistication on the part of the user. Unfortunately, all too often the curve-fitting software is treated as a "black box," which, when fed experimental results, produces (one hopes) a curve that pleasingly fits the data. The black box may also produce various statistical parameters whose intelligibility and utility are more or less obscure. However, a responsible investigator will want to know, at least in broad terms, what the software is doing and why.

6.4.1 ERRORS IN DEPENDENT AND INDEPENDENT VARIABLES

Statistical analysis of binding data must start with a clear appreciation of the distinction between dependent and independent variables in a binding titration. These must be separated unequivocally, in order to carry through a proper analysis. The various linear plots discussed in Chapters 2 and 3 often do not clearly separate dependent and independent variables; the Scatchard plot was noted as an especially salient example. Nonlinear curve fitting that keeps dependent and independent variables separated is much to be preferred.

It should be obvious that the two most important independent variables in a binding experiment will be the amounts (total) of ligand and macromolecule added; of course, there are also other controllable variables such as the pH, salt concentration, and temperature. In cases in which high concentrations of ligand or macromolecule must be used (e.g., for weak binding), or when a cosolvent is added (e.g., ethanol, dimethylsulfoxide, etc.), it may be necessary to allow for volume changes. This will be particularly important when using the familiar molarity concentration scale (the one most natural for interpreting binding isotherms), less so when using molality or mole fractions or mass fractions.

Before starting a titration it is important to determine as accurately as possible the concentrations of macromolecule and ligand in stock solutions; one can then dilute the stocks to get working concentrations. Errors in measuring masses and volumes will need to be determined, and propagated through to calculate or estimate the error in concentrations. Ordinarily, it should be possible to determine the stock concentrations with an error of one or two parts per thousand, with a slightly greater error for the working concentrations due to errors in dilution. There is, however, a particular problem with calculating the concentration of *free* ligand, since this is made by subtracting bound from total ligand. While *total* amounts may be known quite accurately, the *free* amounts may have considerably greater error, due to propagation of error in calculating the amount bound.

The dependent variables in a binding experiment will be quantities such as readings of absorbance or fluorescence intensity, counts of radioactive decays per minute, etc. If one chooses not to follow the model-free method of analysis laid out above, these dependent variables must be related by calculation to the moles of material bound, and then by further calculation converted to the binding density or degree of saturation. Again, it will be necessary to propagate the experimental error in the original readings (along with the errors in volume or mass) through to the final quantity of binding density or degree of saturation. A general feature here is the inconstancy of error in the dependent variable across the titration curve. There is a tendency for this error to increase with the intensity of the signal or response. Consequently, it is necessary to consider how to properly account for the changes in error in the data across the isotherm. This leads to the issue of proper *weighting* of the data, to emphasize the more reliable points in a statistical sense and to discount the less reliable ones. (This is, of course, also an issue with the model-free analysis described above.)

Because of the wide variety of instrumentation used in binding studies, it is not feasible to give here any detailed consideration of experimental error stemming from the measuring instrument. The instrument manufacturer's specifications should be consulted for such details. However, it is usually true that the error in determining the binding density or degree of saturation will be appreciably greater than the error in the independent variables. This is why one often sees only the error in the dependent variable reported in plots of binding isotherms.

6.4.2 COMPUTERIZED FITTING OF DATA

The various linear plots (Scatchard-type, double-reciprocal, half-reciprocal) are quite unsatisfactory for quantitative work. First, in two of these plots (Scatchard-type and half-reciprocal) there is confusion of dependent and independent variables, and hence confusion of errors. Second, curved plots here are capable of multiple interpretations. Remember that the basic model behind the Scatchard-type plot and its relatives assumes only one class of sites. Thus, a curved plot indicates failure of the basic model, but it is not yet possible to say which of several possible, and more complicated, models should be applied. Also, there will often be growth of the error in the dependent variable as the signal strength increases, with the possible result that errors in the dependent variable may conceal curvature in these plots.

Third, due to the dependence of errors on concentration and signal intensity, the error envelope (the two curves bracketing the anticipated spread in the data due to errors) will be such that slopes and intercepts of fitted straight lines will have large errors, making interpretation even more difficult. This is true even when the data are statistically weighted to account for error. Double-reciprocal plots are particularly deplorable, since they tend to place the greatest weight on the least reliable points. Furthermore, the weighting factors can differ across these plots by a factor of 1000 or more. Half-reciprocal plots are better in this respect, but for the least distortion of the data, the best graphical representation is still the nonlinear direct plot of $< r >$ or Y versus log [L]. And for a true thermodynamic isotherm, one or another version of model-free analysis (e.g., that of Bujalowski and Lohman [31,33]), which compares two or more sets of titration data to generate an isotherm, can be recommended. Although this last approach requires an extensive data set, it makes no assumptions about the underlying model of the binding.

6.4.2.1 Nonlinear Curve Fitting

Since the relation between dependent and independent binding variables is nonlinear, the situation calls for nonlinear curve fitting. Considering the wide availability of personal computers and of inexpensive software for such curve fitting, there can be no practical objection to doing this. The following is a brief discussion of the process of curve fitting that is aimed at dispelling some of the obscurity of the fitting process. The discussion is by no means complete, and proper textbooks and reviews on the statistical treatment of data and the algorithms of curve fitting should be consulted for the many details that are necessarily omitted here [38–44].

Suppose that one has in mind a particular binding model, a function B of one or more parameters $\{a_i\}$ (e.g., binding constants, numbers of binding sites) whose values are being sought, along with an experimental variable x that is manipulated (e.g., the ligand concentration). This function will typically not be a linear relation of dependent and independent variable but instead it will be a hyperbola, a sigmoid curve, etc. This dependence is expressed as

$$y = B(x; a_1, a_2, \ldots) \tag{6.12}$$

where y represents the particular value generated by the function B (y might be the degree of binding saturation, for example). As the experiment proceeds, for each value of x, say x_i, there will be generated a corresponding value of y, say y_j. A common scheme for fitting a set of n data points is to manipulate the values of the parameters $\{a_i\}$ so as to minimize a function Φ, the sum of the squares of the vertical distances of the y_i from the curve generated by the current set of $\{a_i\}$:

$$\Phi = \sum_{j=1}^{n} [y_j - B(x_j; a_1, a_2, \ldots)]^2 \tag{6.13}$$

The set of values $\{a_i\}$ that minimizes Φ are then taken as "best estimates" of the parameters in the underlying binding model. So long as the experimental errors are not connected in a systematic way to the x and y values, this is quite satisfactory. However, it is often the case that there *is* a systematic relation between the errors and the x or y values. For example, the error in y may be a constant fraction of the value of y. Then the above relation will tend to place too much reliance on the points with large y values and to ignore those with small y values. Obviously, this could well lead to incorrect estimates of the parameters $\{a_i\}$.

To avoid this, one may wish to weight the data differentially, expressing in some way the degree of confidence in the different data points. To do this, one can assign a weight W_j to the j-th data point, where W_j is a number between 0 and 1 that represents the relative confidence one has in that datum. The fitting of the function B should then involve minimization of a different version of Φ, now expressed as

$$\Phi = \sum_{j=1}^{n} W_j [y_j - B(x_j; a_1, a_2, \ldots)]^2 \tag{6.14}$$

If possible, an empirically determined weighting of the data should be used, which will likely be nonuniform; uniform weighting is not usually realistic (though it *is* convenient, and it may be the most conservative course when there is uncertainty about the distribution of error). For example, tightly grouped replicates might be regarded as more reliable than single points, and so the replicates might be weighted more heavily. It might also be desirable to allow for greater error in points with large y values. Often, however, there is not enough material for replicate titrations to gather empirical measures of the errors or spread in the data. In the absence of good empirical measures one could instead apply some commonly accepted weighting formulas.

A simple scheme to reduce an undesired overreliance on points with large y values is to divide the deviation of each point from the curve by the y value of that point, then to square the result:

$$\Phi = \sum_{j=1}^{n} \left[\frac{y_j - B(x; a_1, a_2, \ldots)}{y_j} \right]^2 \tag{6.15}$$

Another common weighting scheme is to use a weight W_j for the j-th measurement that is inversely proportional to the (estimated) variance of the measurement:

$$W_j = \frac{1}{\sigma_j^2} \tag{6.16}$$

The variance σ_j at a point y_j is defined as

$$\sigma_j = \sum_{i=1}^{m} \left(\frac{[y_{ij} - <y_j>]^2}{m} \right)^{1/2} \tag{6.17}$$

and $<y_j>$ represents the mean result of m repeated measurements at point j. Weighting points according to the inverse of their variance may, however, lead to overreliance on small measurements, which receive greater weight because their variance is predicted to be small. Consequently, this weighting scheme can cause unreasonably high estimates of a high-affinity K (for a multiclass model), particularly if the high-affinity class of sites is greatly outnumbered by the lower-affinity sites.

Other models might be used for the variance, and depending on the computer software, there might be a choice among several models. When in doubt, however, it is probably safest to weight the data equally, and to use differential weighting only when there is a clear relation between the measurements and their uncertainties. One caution: if the data are transformed from one format to another (as when forming a Hill plot from data on the degree of binding saturation), then one should be especially careful to apply the correct weighting to the transformed data. For example, suppose that the data are weighted according to $W_j = \sigma_j^{-2}$ and the measurements y_j are to be transformed to a new format y_j'. Then the variances may be transformed according to

$$(\sigma_j')^2 = \left(\frac{\partial y_j'}{\partial y_j} \right)^2 (\sigma_j)^2 \tag{6.18}$$

so long as the errors are small and the transformation from y_j to y_j' is continuous and bounded, and so are all its derivatives. Suppose that the original y_j were measurements of the degree of saturation Y, and that the data are transformed to Hill plot format, with y_j' equal to $\ln[Y_j/(1 - Y_j)]$. Then if the original data were uniformly weighted (all σ_j^{-2} equal to 1) the weighting in the Hill plot should now be according to $W_j' = Y_j^2 (1 - Y_j)^2$.

At this point it is necessary to consider corrections for nonspecific binding. This is not merely a simple correction for ligand lost to centrifuge tubes, glass fibers, etc., but is also an accounting for ligand bound to (apparently) nonsaturable sites on receptors (e.g., partitioning of a hydrophobic ligand into a lipid bilayer around an embedded protein receptor). The simplest approach is to subtract nonspecific binding from the total binding (after first correcting for "lost ligand" on glassware, etc.). This can produce small values of "specifically bound" ligand that may receive an inordinately high statistical weight if $W_j = \sigma_j^{-2}$ is used. Alternatively, it is possible to use a binding isotherm that includes a class of numerous, low-affinity sites in addition to much less numerous high-affinity sites. This stratagem may allow later recovery of useful information about secondary binding sites on the macromolecule.

A good software package for data analysis will report several statistical quantities and generate different graphical displays of the results. First, it should, of course, show a plot of the data overlaid with the fitted function B. This will permit an immediate visual inspection of the general goodness of fit of the function. More objectively, it should give the root mean square (RMS) error, the magnitude of the average deviation of a typical data point from the curve. An inspection of the plot may also hint at trends in the fitting of B to the data, regions where data lie above or below the curve. This can be measured objectively by calculating the residuals, the distance of a datum from the curve. Thus, a second desirable feature is that the software should generate a plot of the residuals as a function of the $\{x_j\}$ (see Figure 6.3).

If the data points scatter randomly above and below the model curve, this can be taken as an indication of a good fit of the model to the data. The signs of the corresponding residuals should be distributed randomly; there should not be large groups of adjacent residuals with the same sign. There are tables available of the expected number of randomly occurring "runs" for data sets of different sizes. If the actual number of observed runs exceeds, or is much less than, the tabulated figure for that size of data set, then the chosen model may not be appropriate and alternatives should be considered. This runs test does break down, however, if the magnitude of the residuals is very small. For further details on this test see Straume and Johnson [43].

What about "error bars" on the fitted model parameters? With nonlinear functions the computed standard errors in the fitted parameters are based on a number of simplifying assumptions that generally cause the reported values to be underestimates of the true uncertainty in those values. Thus, the errors in fitted values of n and K from a nonlinear isotherm equation (like the Langmuir isotherm) should not be taken too seriously.

It is possible that more than one model will give a reasonable fit to the data, i.e., have a low RMS error. Then careful reporting calls for a comparison of the fit of one model with another. A simple first test is a comparison of the RMS error for each model. For models with the same number of parameters and with the same weighting scheme for the data, this settles the matter; one should prefer the model yielding the lower RMS error, since its curve lies closer to the data.

When comparing models with different numbers of parameters, things become more difficult. Introducing more parameters into the function B almost always improves the fit, since the function B is now more flexible. But are the extra parameters justifiable in an objective sense? Do they improve the fit more than would be expected by random chance? The F test is an objective statistical comparison that is often useful in these situations.

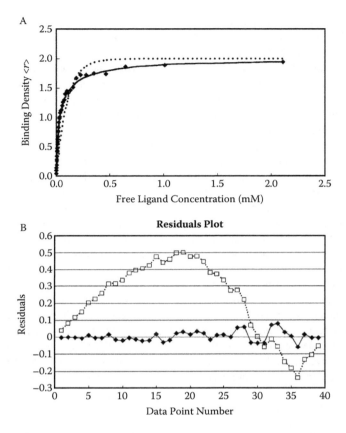

FIGURE 6.3 Detecting randomness of error in fitting isotherm data. A. Modeling a system with two unequal, independent sites per macromolecule: $k_1 = 1.0 \times 10^5$ M^{-1} and $k_2 = 1.0 \times 10^4$ M^{-1}. Random error of up to $\pm 5\%$ in binding density ($\blacklozenge$) superimposed on the true isotherm (solid line). For comparison a hypothetical exponential isotherm (dotted line) is shown, generated from $<r> = 2 \times (1 - \exp(-k_e \, [\text{L}]))$, with $k_e = 2.19 \times 10^4$ M^{-1}. B. Residuals plot, comparing the highly nonrandom residuals from the exponential isotherm ($\square$, two runs) versus the satisfactorily random residuals from the true compound hyperbolic isotherm ($\blacklozenge$, 17 runs, within the expected range of 16 to 26 runs).

Suppose that model 1 has a residual sum of squares SS_1 and a certain number of degrees of statistical freedom df_1 (the statistical degree of freedom is defined as the number of data points minus the number of parameters). Model 2 is more complex, with more parameters. It has a sum of squares SS_2 and degrees of freedom df_2, which differs from df_1 since model 2 has a different number of parameters. The F value is defined as

$$F = \frac{(SS_1 - SS_2)^2/(df_1 - df_2)}{SS_2/df_2} \tag{6.19}$$

Standard tables of a quantity p tabulated as a function of F, $(df_1 - df_2)$, and df_2 are available. The smaller the value of p, the more likely it is that the more complex model is to be preferred.

There is, of course, the overall plausibility of the model. Unusual values of the fitted parameters may simply not make physical sense—one should strongly doubt a fitted model giving negative values for the binding constant! Possibly, the values of the parameters would contradict information from other sources, for example, kinetic studies that reveal multiple classes of binding sites (slow versus fast binding) where only one is indicated by the fit using equilibrium data. It is best to avoid premature assumption of a particular binding model and to use the most general analysis possible when seeking to measure binding affinity. An apparent binding constant can be easily obtained by using a plot of Y as a function of log [L]; see Chapter 2. This may suffice for many applications, e.g., broad comparisons of a set of enzyme inhibitors or receptor antagonists. Once K is found one can then go on to determine the stoichiometry. This is where model assumptions become particularly important (e.g., multiple classes of sites versus negative cooperativity among identical sites), and extra-thermodynamic data (e.g., structural determinations by NMR or by X-ray diffraction, or binding kinetics) become especially useful in choosing among models.

The desirability of confirming results by using two independent experimental techniques was noted earlier. These methods may, of course, involve different background corrections, different weighting schemes for the data, and so on. The software package for data analysis should be capable of handling data coming in different formats, without undue strain on the investigator for format conversion. There are several other desirable qualities for a good data analysis software package, which is summarized in the list below.

Permits nonlinear fitting schemes

Allows a choice of data weighting schemes

Can include a correction for "background," and for nonspecific and non-saturable binding

Automatically converts raw data to a "polished" form suitable for professional publication

Easy to learn

Fast in operation

Allows data from different experimental methods to be compared and combined

Accepts multiple file formats (including common formats such as comma-delimited, tabbed, etc., and including those from common spreadsheet software)

Permits online editing of data files, to compare effects of eliminating outliers, etc.

Accepts results from different experimental runs, overlays results

Multiple binding models available

Allows for fitting of nonspecific/nonsaturable binding

Reports useful statistics: not just values of parameters that were "fitted," but the error in those values

Compares model choices statistically (e.g., through F-test)

The use of computer methods for data analysis does not absolve the experimenter from proper design and execution of the experiments, or from applying these aids critically and responsibly. The careful experimenter should choose an experimental method that will generate a maximum response that is much larger than the error in individual data points. Measurements should be taken over a range of concentrations that at least cover the central portion of the isotherm, and which are preferably extended to as high and low binding densities as may be feasible [22,24,25]. A limited set of measurements that covers less than half the range of binding density can produce highly misleading values for the affinity and stoichiometry. With proper planning, the range and spacing can be optimized for the independent variable, whether it is the concentration of free ligand or of macromolecule. The scheme for weighting of the data should also be carefully considered. And finally, unreplicated data should always be considered suspect; independent duplicates or (better) triplicates of each measurement should be made.

REFERENCES

1. Cantor, C.R. and Schimmel, P.R., *Biophysical Chemistry, Part II: Techniques for the Study of Biological Structure and Function.* San Francisco: WH Freeman, 1980.
2. Campbell, I.D. and Dwek, R.A., *Biological Spectroscopy.* Menlo Park, CA: Benjamin/ Cummings, 1984.
3. Connors, K.A., *Binding Constants: The Measurement of Molecular Complex Stability.* New York: Wiley-Interscience, 1987.
4. Hulme, E.C., *Receptor-Ligand Interactions: A Practical Approach.* Oxford: IRL Press, 1992.
5. Klotz, I.M., *Ligand-Receptor Energetics: A Guide for the Perplexed.* New York: Wiley-Interscience, 1997.
6. Winzor, D.J. and Sawyer, W.H., *Quantitative Characterization of Ligand Binding.* New York: Wiley-Liss, 1965.
7. Honig, B. and Nicholls, A., Classical electrostatics in biology and chemistry, *Science* 268, 1144, 1995.
8. Norberg, J., Association of protein-DNA recognition complexes: Electrostatic and nonelectrostatic effects, *Arch. Biochem. Biophys.* 410, 48, 2003.
9. Record, M.T. Jr., Anderson, C.F., and Lohman, T.L., Thermodynamic analyis of ion effects on the binding and conformational equilibria of proteins and nucleic acids: The roles of ion association or release, screening, and ion effects on water activity, *Q. Rev. Biophys.* 11, 103, 1978.
10. Anderson, C.F. and Record, M.T. Jr., Salt-nucleic acid interactions, *Annu. Rev. Phys. Chem.* 46, 657, 1995.
11. Record, M.T. Jr., Zhang, W., and Anderson, C.F., Analysis of effects of salts and uncharged solutes on protein and nucleic acid equilibria and processes: A practical guide to recognizing and interpreting polyelectrolyte effects, Hofmeister effects and osmotic effects of salts, *Adv. Protein Chem.* 51, 281, 1998.
12. Record, M.T. Jr., Ha, J.-H., and Fisher, M.A., Analysis of equilibrium and kinetic measurements to determine thermodynamic origins of stability and specificity and mechanism of formation of site-specific complexes between proteins and helical DNA, *Meth. Enzymol.* 208, 291, 1991.
13. Lewis, G.N. and Randall, M., *Thermodynamics,* 2nd ed., revised by Pitzer, K.S. and Brewer, L. New York: McGraw-Hill, 1961.

14. Butler, J.N., *Ionic Equilibrium: A Mathematical Approach*. Reading, MA: Addison-Wesley, 1964.
15. Pailthorpe, B.A., Mitchell, D.J., and Ninham, B.W., Ion-solvent interactions and the activity coefficients of real electrolyte solutions, *J. Chem. Soc. Faraday Trans.* 2 (80), 115, 1984.
16. Khoo, K.H., Activity coefficients in mixed-electrolyte solutions, *J. Chem. Soc. Faraday Trans.* 1 (82), 1, 1986.
17. Fulton, A.B., How crowded is the cytoplasm? *Cell* 30, 345, 1982.
18. Zimmerman, S.B. and Minton, A.P., Macromolecular crowding: Biochemical, biophysical, and physiological consequences, *Annu. Rev. Biophys. Biomol. Struct.* 22, 27, 1993.
19. Minton, A.P., Influence of excluded volume upon macromolecular structure and associations in 'crowded' media, *Curr. Opin. Biotechnol.* 8, 65, 1997.
20. Minton, A.P., Molecular crowding: Analysis of effects of high concentrations of inert cosolutes on biochemical equilibria and rates in terms of volume exclusion, *Meth. Enzymol.* 295,127, 1998.
21. Weber, G., The binding of small molecules to proteins, in *Molecular Biophysics*, Pullman, B. and Weissbluth, M., Eds. New York: Academic Press, 1965, p. 369.
22. Deranleau, D.A., Theory of the measurement of weak molecular complexes. I. General considerations, *J. Am. Chem. Soc.* 91, 4044, 1969a.
23. Deranleau, D.A., Theory of the measurement of weak molecular complexes. II. Consequences of multiple equilibria, *J. Am. Chem. Soc.* 91, 4050, 1969b.
24. Bowser, M.T. and Chen, D.D.Y., Monte Carlo simulation of error propagation in the determination of binding constants from rectangular hyperbolae. 1. Ligand concentration range and binding constant, *J. Phys. Chem. A* 102, 8063, 1998.
25. Bowser, M.T. and Chen, D.D.Y., Monte Carlo simulation of error propagation in the determination of binding constants from rectangular hyperbolae. 2. Effect of the maximum-response range, *J. Phys. Chem. A* 103, 197, 1999.
26. Cleland, W.W., The statistical analysis of enzyme kinetic data, *Adv. Enzymol.* 29, 1, 1967.
27. Klotz, I.M., Numbers of receptor sites from Scatchard graphs: Facts and fantasies, *Science* 217, 1247, 1982.
28. Halfman, C.J. and Nishida, T., Method for measuring the binding of small molecules to proteins from binding-induced alterations of physical-chemical properties, *Biochemistry* 18, 3493, 1972.
29. Schütz, H. et al., Design and data analysis of DNA binding isotherms, *Studia Biophysica* 104, 23, 1984.
30. Schwarz, G., Stankowski, S., and Rizzo, V., Thermodynamic analysis of incorporation and aggregation in a membrane: Application to the pore-forming peptide alamethicin, *Biochim. Biophys. Acta* 861, 141, 1987.
31. Bujalowski, W. and Lohman, T.M., A general method of analysis of ligand-macromolecule equilibria using a spectroscopic signal from the ligand to monitor binding. Application to *Escherichia coli* single-strand binding protein-nucleic acid interactions, *Biochemistry* 26, 3099, 987.
32. Chatelier, R.C. and Sawyer, W.H., Isoparametric analysis of binding and partitioning processes, *J. Biochem. Biophys. Meth.* 15, 49, 1987.
33. Lohman, T.L. and Bujalowski, W., Thermodynamic methods for model-independent determination of equilibrium binding isotherms for protein-DNA interactions: Spectroscopic approaches to monitor binding, *Meth. Enzymol.* 208, 258, 1991.
34. Schwarz, G., A universal thermodynamic approach to analyze biomolecular binding experiments, *Biophys. Chem.* 86, 119, 2000.

35. Bujalowski, W., and Klonowska, M.M., Negative cooperativity in the binding of nucleotides to *Escherichia coli* replicative helicase DnaB protein. Interactions with fluorescent nucleotide analogs, *Biochemistry* 32, 5888, 1993.
36. Jezewska, M.J. and Bujalowski, W., A general method of analysis of ligand binding to competing macromolecules using the spectroscopic signal originating from a reference macromolecule. Application to *Escherichia coli* replicative helicase DnaB protein-nucleic acid interactions, *Biochemistry* 35, 2117, 1996.
37. Raguin, O., Gruaz-Guyon, A., and Barbet, J., Equilibrium expert: An add-in to Microsoft Excel for multiple binding equilibrium simulations and parameter estimations, *Anal. Biochem.* 310, 1, 2002.
38. Lundeen, J.E. and Gordon, J.H., Computer analysis of binding data, in *Receptor Binding in Drug Research,* O'Brien, R.A., Ed. New York: Marcel Dekker, 1986, p. 31.
39. Press, W.H. et al., *Numerical Recipes: The Art of Scientific Computing.* Cambridge, UK: Cambridge University Press, 1986.
40. Bevington, P.R. and Robinson, D.K., *Data Reduction and Error Analysis for the Physical Sciences,* 2nd ed. New York: McGraw-Hill, 1982.
41. Di Cera, E., Use of weighting functions in data fitting, *Meth. Enzymol.* 210, 68, 1992.
42. Johnson, M.L., Analysis of ligand-binding data with experimental uncertainties in independent variables, *Meth. Enzymol.* 210, 106, 1992.
43. Straume, M. and Johnson, M.L., Analysis of residuals: Criteria for determining goodness-of-fit, *Meth. Enzymol.* 210, 87, 1992.
44. Myung, J.I. and Pitt, M.A., Model comparison methods, *Meth. Enzymol.* 383, 351, 2004.

Index

9 780367 388324